I0489296

Status of Groundwater Quality in the Coastal Los Angeles Basin, 2006: California GAMA Priority Basin Project

By Dara Goldrath, Miranda S. Fram, Michael Land, and Kenneth Belitz

A product of the California Groundwater Ambient Monitoring and Assessment (GAMA) Program

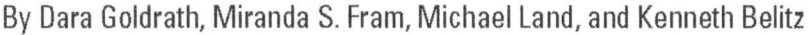

Prepared in cooperation with the California State Water Resources Control Board

Scientific Investigations Report 2012–5048

U.S. Department of the Interior
U.S. Geological Survey

U.S. Department of the Interior
KEN SALAZAR, Secretary

U.S. Geological Survey
Marcia K. McNutt, Director

U.S. Geological Survey, Reston, Virginia: 2012

For more information on the USGS—the Federal source for science about the Earth, its natural and living resources, natural hazards, and the environment, visit http://www.usgs.gov or call 1–888–ASK–USGS.

For an overview of USGS information products, including maps, imagery, and publications, visit http://www.usgs.gov/pubprod

To order this and other USGS information products, visit http://store.usgs.gov

Contents

Figures

Figures—Continued

Tables

Conversion Factors, Datums, and Abbreviations and Acronyms

Conversion Factors

Inch/foot/mile to International System of Units (SI)

Multiply	By	To obtain
Length		
inch (in.)	2.54	centimeter (cm)
inch (in.)	25.4	millimeter (mm)
foot (ft)	0.3048	meter (m)
mile (mi)	1.609	kilometer (km)
Area		
square foot (ft^2)	0.09290	square meter (m^2)
square mile (mi^2)	2.590	square kilometer (km^2)
Radioactivity		
picocurie per liter (pCi/L)	0.037	becquerel per liter (Bq/L)

Temperature in degrees Celsius (°C) may be converted to degrees Fahrenheit (°F) as follows:

$$°F=(1.8×°C)+32$$

Temperature in degrees Fahrenheit (°F) may be converted to degrees Celsius (°C) as follows:

$$°C=(°F-32)/1.8$$

Specific conductance is given in microsiemens per centimeter at 25 degrees Celsius (μS/cm at 25 °C).

Concentrations of chemical constituents in water are given either in milligrams per liter (mg/L) or micrograms per liter (μg/L) or nanograms per liter (ng/L). One milligram per liter is equivalent to 1 part per million (ppm); 1 microgram per liter is equivalent to 1 part per billion (ppb); 1 nanogram per liter (ng/L) is equivalent to 1 part per trillion (ppt); 1 per mil is equivalent to 1 part per thousand.

Datums

Vertical coordinate information is referenced to the North American Vertical Datum of 1988 (NAVD 88).

Horizontal coordinate information is referenced to the North American Datum of 1983 (NAD 83).

Conversion Factors, Datums, and Abbreviations and Acronyms—Continued

Abbreviations and Acronyms

AL-US	U.S. Environmental Protection Agency action level
E	estimated or having a higher degree of uncertainty
CB	Central Basin study area
DA	direct-assessment well
CLAB	Coastal Los Angeles Basin study unit
GAMA	Groundwater Ambient Monitoring and Assessment Program
HAL-US	U.S. Environmental Protection Agency lifetime health advisory level
HBSL	Health-based screening levels
LRL	laboratory reporting level
LSD	land-surface datum
LT-MDL	long-term method detection level
MCL	maximum contaminant level
MCL-CA	California Department of Public Health maximum contaminant level
MCL-US	U.S. Environmental Protection Agency maximum contaminant level
MDL	method detection limit
MRL	minimum reporting level
OC	Orange County Coastal Plain study area
na	no data available
NAVD 88	North American Vertical Datum of 1988
nc	no significant correlation
nd	constituent not detected
ns	not sampled
NL-CA	California Department of Public Health notification level
QC	quality control
RSD5-US	U.S. Environmental Protection Agency risk-specific dose at a risk factor of 10^{-5}
SMCL-CA	California Department of Public Health secondary maximum contaminant level
SMCL-US	U.S. Environmental Protection Agency secondary maximum contaminant level
TEAP	Terminal Electron Acceptor Processes
TT-US	treatment technique levels
U	understanding well
WB	West Coast Basin study area

Organizations

CDPH	California Department of Public Health (Department of Health Services prior to July 1, 2007)
CDPR	California Department of Pesticide Regulation
CDWR	California Department of Water Resources
LLNL	Lawrence Livermore National Laboratory
NAWQA	National Water-Quality Assessment (USGS) Program
SWRCB	State Water Resources Control Board (California)
USEPA	U.S. Environmental Protection Agency
USGS	U.S. Geological Survey

Conversion Factors, Datums, and Abbreviations and Acronyms—Continued

Selected Chemical Names

Ammonia-N	ammonia as nitrogen
DBCP	1,2-dibromo-3-chloropropane
DEHP	bis(2-ethylhexyl) phthalate
DOC	dissolved organic carbon
EDB	1,2-dibromomethane (ethylene dibromide)
MTBE	methyl *tert*-butyl ether
NDMA	*N*-nitrosodimethylamine
Nitrate-N	nitrate as nitrogen
Nitrite-N	nitrite as nitrogen
PCE	tetrachloroethene
TCE	trichloroethene
1,2,3-TCP	1,2,3-trichloropropane
TDS	total dissolved solids
THM	trihalomethane
VOC	volatile organic compound

Units of Measure

a	annum (year)
cm^3 STP g^{-1}	cubic centimeters at standard temperature and pressure per gram
δ	delta notation; the ratio of a heavier isotope to the more common lighter isotope of an element, relative to a standard reference material, expressed as per mil
gpm	gallons per minute
kg	kilogram
L	liter
L/min	liters per minute
Ma	Megaannum (million year)
mg/L	milligrams per liter (parts per million)
mL	milliliter
mm	millimeter
µg/L	micrograms per liter (parts per billion)
µS/cm	microsiemens per centimeter
per mil	parts per thousand
pmc	percent modern carbon
ppm	parts per million
TU	tritium unit
yr	year
>	greater than
<	less than
%	percent

Status of Groundwater Quality in the Coastal Los Angeles Basin, 2006: California GAMA Priority Basin Project

By Dara Goldrath, Miranda S. Fram, Michael Land, and Kenneth Belitz

Abstract

Groundwater quality in the approximately 860-square-mile (2,227-square-kilometer) Coastal Los Angeles Basin study unit (CLAB) was investigated as part of the Priority Basin Project of the Groundwater Ambient Monitoring and Assessment (GAMA) Program. The study area is located in southern California in Los Angeles and Orange Counties. The GAMA Priority Basin Project is being conducted by the California State Water Resources Control Board in collaboration with the U.S. Geological Survey (USGS) and the Lawrence Livermore National Laboratory.

The GAMA CLAB study was designed to provide a spatially unbiased assessment of the quality of untreated (raw) groundwater in the primary aquifer system. The assessment is based on water-quality and ancillary data collected in 2006 by the USGS from 69 wells and on water-quality data from the California Department of Public Health (CDPH) database. The primary aquifer system was defined by the depth interval of the wells listed in the CDPH database for the CLAB study unit. The quality of groundwater in the primary aquifer system may be different from that in the shallower or deeper water-bearing zones; shallow groundwater may be more vulnerable to surficial contamination.

This study assesses the status of the current quality of the groundwater resource by using data from samples analyzed for volatile organic compounds (VOCs), pesticides, and naturally occurring inorganic constituents, such as major ions and trace elements. This status assessment is intended to characterize the quality of groundwater resources in the primary aquifer system of the CLAB study unit, not the treated drinking water delivered to consumers by water purveyors.

Relative-concentrations (sample concentration divided by the health- or aesthetic-based benchmark concentration) were used for evaluating groundwater quality for those constituents that have Federal and (or) California regulatory or non-regulatory benchmarks for drinking-water quality. A relative-concentration greater than ($>$) 1.0 indicates a concentration greater than a benchmark, and a relative-concentration less than or equal to ($\leq$) 1.0 indicates a concentration equal to or less than a benchmark. Relative-concentrations of organic and special-interest constituents [perchlorate, N-nitrosodimethylamine (NDMA), 1,2,3-trichloropropane (1,2,3-TCP), and 1,4-dioxane] were classified as "high" (relative-concentration$>$1.0), "moderate" (0.1$<$relative-concentration$\leq$1.0), or "low" (relative-concentration$\leq$0.1. Relative-concentrations of inorganic constituent were classified as "high" (relative-concentration$>$1.0), "moderate" (0.5$<$relative-concentration$\leq$1.0), or "low" (relative-concentration$\leq$0.5).

Aquifer-scale proportion was used as the primary metric in the *status assessment* for evaluating regional-scale groundwater quality. High aquifer-scale proportion is defined as the percentage of the area of the primary aquifer system with a relative-concentration greater than 1.0 for a particular constituent or class of constituents; percentage is based on an areal rather than a volumetric basis. Moderate and low aquifer-scale proportions were defined as the percentage of the primary aquifer system with moderate and low relative-concentrations, respectively. Two statistical approaches—grid-based and spatially weighted—were used to evaluate aquifer-scale proportions for individual constituents and classes of constituents. Grid-based and spatially weighted estimates were comparable in the CLAB study unit (within 90-percent confidence intervals).

Inorganic constituents with human-health benchmarks were detected at high relative-concentrations in 5.6 percent of the primary aquifer system and moderate in 26 percent. High aquifer-scale proportion of inorganic constituents primarily reflected high aquifer-scale proportions of arsenic (1.9 percent), nitrate (1.9 percent), and uranium (1.2 percent). Inorganic constituents with secondary maximum contaminant levels (SMCL) were detected at high relative-concentrations in 18 percent of the primary aquifer system and moderate in 47 percent. The constituents present at high relative-concentrations included total dissolved solids (1.9 percent), manganese (15 percent), and iron (9.4 percent).

Relative-concentrations of organic constituents (one or more) were high in 3.7 percent, and moderate in 13 percent, of the primary aquifer system. The high aquifer-scale proportion of organic constituents primarily reflected high aquifer-scale proportions of solvents, including trichloroethene (TCE; 1.7 percent), perchloroethene (PCE; 1.1 percent), and carbon tetrachloride (1.0 percent). Of the 204 organic constituents analyzed, 44 constituents were detected. Eleven organic constituents had detection frequencies of greater than 10 percent: the trihalomethanes chloroform and bromodichloromethane, the solvents TCE, PCE, *cis*-1,2-dichloroethene, and 1,1-dichloroethene, the herbicides atrazine, simazine, prometon, and tebuthiuron, and the gasoline additive methyl *tert*-butyl ether (MTBE). Most detections were at low relative-concentrations.

The special-interest constituent perchlorate was detected at high relative-concentrations in 0.5 percent of the primary aquifer system, and at moderate relative-concentrations in 35 percent. The special-interest constituent 1,4-dioxane was detected at high relative-concentrations, but an insufficient number of samples was analyzed to provide a representative estimate of aquifer-scale proportion.

Introduction

To assess the quality of ambient groundwater in aquifers used for drinking-water supply and to establish a baseline groundwater-quality monitoring program, the State Water Resources Control Board (SWRCB), in collaboration with the U.S. Geological Survey (USGS) and Lawrence Livermore National Laboratory (LLNL), implemented the Groundwater Ambient Monitoring and Assessment (GAMA) Program (website at http://www.waterboards.ca.gov/gama/). The statewide GAMA Program currently consists of three projects: (1) the GAMA Priority Basin Project, conducted by the USGS (website at http://ca.water.usgs.gov/gama/); (2) the GAMA Domestic Well Project, conducted by the SWRCB; and (3) the GAMA Special Studies, conducted by LLNL. On a statewide basis, the Priority Basin Project focused on the primary aquifer system, typically the deep portion of the groundwater resource, and the SWRCB Domestic Well Project generally focused on the shallow aquifer system. The deeper aquifers may be at less risk of contamination than the shallow wells, such as private domestic and environmental monitoring wells, which are closer to surficial sources of contamination. As a result, concentrations of contaminants, such as volatile organic compounds (VOCs) and nitrate, in wells screened in the deep aquifers may be lower than concentrations of constituents in shallow wells (Kulongoski and others, 2010; Landon and others, 2010).

The SWRCB initiated the GAMA Program in 2000 in response to Legislative mandates (State of California, 1999,

2001a, Supplemental Report of the 1999 Budget Act 1999–00 Fiscal Year). The GAMA Priority Basin Project was initiated in response to the Groundwater Quality Monitoring Act of 2001 (State of California, 2001b) {Sections 10780–10782.3 of the California Water Code, Assembly Bill 599} to assess and monitor the quality of groundwater in California. The GAMA Priority Basin Project is a comprehensive assessment of statewide groundwater quality designed to help better understand and identify risks to groundwater resources and to increase the availability of information about groundwater quality to the public. For the Priority Basin Project, the USGS, in collaboration with the SWRCB, developed a monitoring plan to assess groundwater basins through direct sampling of groundwater and other statistically reliable sampling approaches (Belitz and others, 2003; State Water Resources Control Board, 2003). Additional partners in the GAMA Priority Basin Project include the California Department of Public Health (CDPH), the California Department of Pesticide Regulation (CDPR), the California Department of Water Resources (CDWR), and local water agencies and well owners (Kulongoski and Belitz, 2004).

The range of hydrologic, geologic, and climatic conditions that exist in California should be considered in an assessment of groundwater quality. Belitz and others (2003) partitioned the State into 10 hydrogeologic provinces, each with distinctive hydrologic, geologic, and climatic characteristics (fig. 1). All these hydrogeologic provinces include groundwater basins and subbasins designated by the CDWR (California Department of Water Resources, 2003). Groundwater basins generally consist of relatively permeable, unconsolidated deposits of alluvial origin. Eighty percent of California's approximately 16,000 public-supply wells are in designated groundwater basins. Groundwater basins and subbasins were prioritized for sampling on the basis of the number of public-supply wells, with secondary consideration given to the municipal population served, the volume of agricultural pumping, the number of historically leaking underground fuel tanks, and the number of square-mile sections having registered pesticide applications (Belitz and others, 2003). The 116 priority basins and additional areas outside defined groundwater basins were grouped into 35 study units, which include approximately 95 percent of public-supply wells in California. The Coastal Los Angeles Basin study unit is composed of five contiguous groundwater basins in the Transverse Ranges and selected Peninsular Ranges hydrogeologic province (fig. 1).

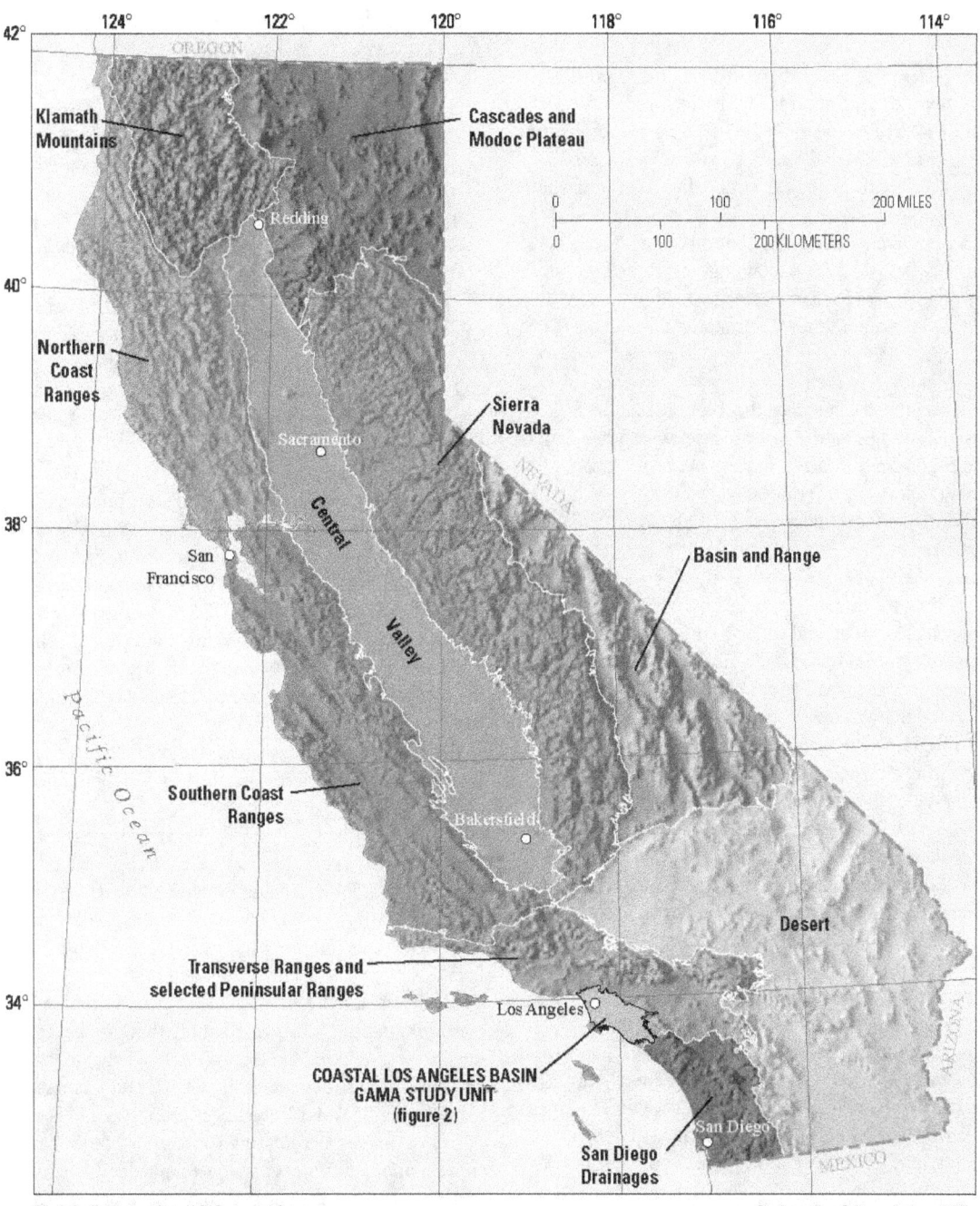

Figure 1.   Location of the Coastal Los Angeles Basin California Groundwater Ambient Monitoring and Assessment (GAMA) study unit, 2006, and California hydrogeologic provinces (modified from Belitz and others, 2003).

Purpose and Scope

The purposes of this report are to provide a (1) *study unit description:* description of the hydrogeologic setting of the Coastal Los Angeles Basin GAMA study unit (hereinafter referred to as the CLAB study unit), (2) *status assessment:* assessment of the status of the current quality of groundwater in the primary aquifer system in the CLAB study unit, and (3) *compilation of ancillary data:* compilation of data for selected factors that may be useful for explaining water quality.

Water-quality data for samples collected by the USGS for the GAMA Program in the CLAB study unit and details of sample collection, analysis, and quality-assurance procedures for the CLAB study unit are reported by Mathany and others (2008). Utilizing those same data, this report describes methods used in designing the sampling network, identifying CDPH data for use in the *status assessment*, estimating aquifer-scale proportions of relative-concentrations, and assessing the status of groundwater quality by statistical and graphical approaches.

The *status assessment* includes evaluation of water-quality data for 69 wells sampled by the USGS, 55 of which were selected for spatial coverage of one well per grid cell (hereinafter referred to as USGS-grid wells), across the CLAB study unit. Water-quality data from the CDPH database were used to supplement data collected by the USGS for the GAMA Program. The resulting set of water-quality data from USGS-grid wells and CDPH wells was considered to be representative of the primary aquifer system in the CLAB study unit; the primary aquifer system is defined by the depth of the screened or perforated intervals of the wells listed in the CDPH database for the CLAB study unit. GAMA *status assessments* are designed to provide a statistically robust characterization of groundwater quality in the primary aquifer system at the basin-scale (Belitz and others, 2003, 2010). The statistically robust design also allows basins to be compared and results to be synthesized regionally and statewide.

To provide context, the water-quality data discussed in this report were compared to California and Federal regulatory and non-regulatory benchmarks for drinking water. The assessments in this report are intended to characterize the quality of untreated groundwater resources in the primary aquifer system within the study unit, not the drinking water delivered to consumers by water purveyors. This study does not attempt to evaluate the quality of water delivered to consumers; after withdrawal from the ground, water typically is treated, disinfected, and (or) blended with other waters to maintain acceptable water quality. Regulatory benchmarks apply to drinking water that is delivered to the consumer, not to untreated groundwater.

Hydrogeologic Setting of the Coastal Los Angeles Basin Study Unit

The CLAB study unit lies at the southwest end of the Transverse Ranges and selected Peninsular Ranges hydrogeologic province described by Belitz and others (2003) and includes five groundwater basins (fig. 2): Santa Monica, Hollywood, West Coast, Central Plain, and the Orange County Coastal Plain (California Department of Water Resources, 2003). The five study areas of CLAB—the Santa Monica Basin, the Hollywood Basin, the West Coast Basin (WB), the Central Basin (CB), and the Orange County Coastal Plain (OC)—generally correspond with the boundaries of the five CDWR-defined groundwater basins and cover an area of approximately 860 square miles (mi²) in Los Angeles and Orange Counties, California.

The area encompassed by the CLAB study unit has been the subject of many hydrogeologic investigations. This report contains a brief summary of the hydrologic setting of the CLAB study unit; more detailed descriptions may be found in other publications. The OC study area corresponds to the Coastal Basin subunit of the USGS National Water-Quality Assessment (NAWQA) Program's Santa Ana Basin study unit (Belitz and others, 2004), and the hydrogeologic setting of the study area was described by Herndon and others (1997) and Hamlin and others (2002). The hydrogeologic settings of the WB and CB study areas and of the Santa Monica and Hollywood basins were described by Reichard and others (2003).

The topography of the CLAB study unit is relatively flat. The study unit is bounded on the north by the Santa Monica Mountains and the Elysian, Repetto, Merced, Puente, and Chino Hills (fig. 2). It is bordered on the east by the Santa Ana Mountains, on the south by the San Joaquin Hills and the Pacific Ocean (San Pedro Bay), and on the west by the Palos Verdes Hills and the Pacific Ocean (Santa Monica Bay). The major drainages of the CLAB study unit are the Los Angeles, the San Gabriel, and the Santa Ana Rivers, all of which have headwaters outside of the CLAB study unit (California Department of Water Resources, 2003).

The main water-bearing formations within the CLAB study unit occur in unconsolidated and semi-consolidated marine and alluvial sediments of Quaternary and late-Tertiary ages (Holocene, Pleistocene, and Pliocene epochs) (fig. 3). Periodic transgressions of the sea and alluvium derived from weathering and erosion of the rocks in the surrounding mountains have filled the Coastal Los Angeles Basin with deposits of various thicknesses that consist of sand, gravel, and conglomerate with some silt and clay beds (California Department of Water Resources, 2003).

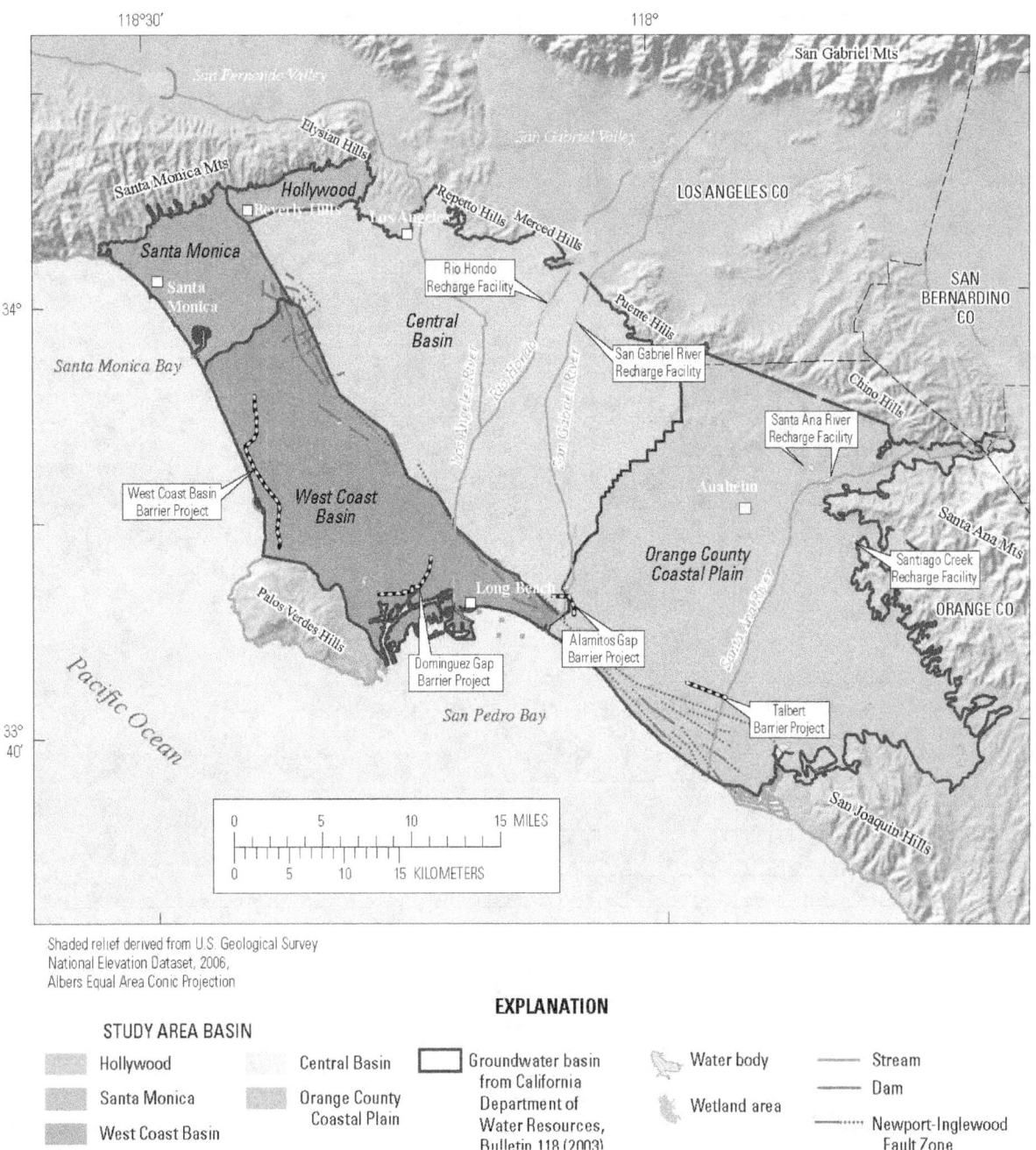

Shaded relief derived from U.S. Geological Survey
National Elevation Dataset, 2006,
Albers Equal Area Conic Projection

EXPLANATION

STUDY AREA BASIN

Hollywood

Santa Monica

West Coast Basin

Central Basin

Orange County
Coastal Plain

Groundwater basin
from California
Department of
Water Resources,
Bulletin 118 (2003)

Water body

Wetland area

Stream

Dam

Newport-Inglewood
Fault Zone

Figure 2. Geographic features of the Coastal Los Angeles Basin study unit, 2006, California GAMA Priority Basin Project.

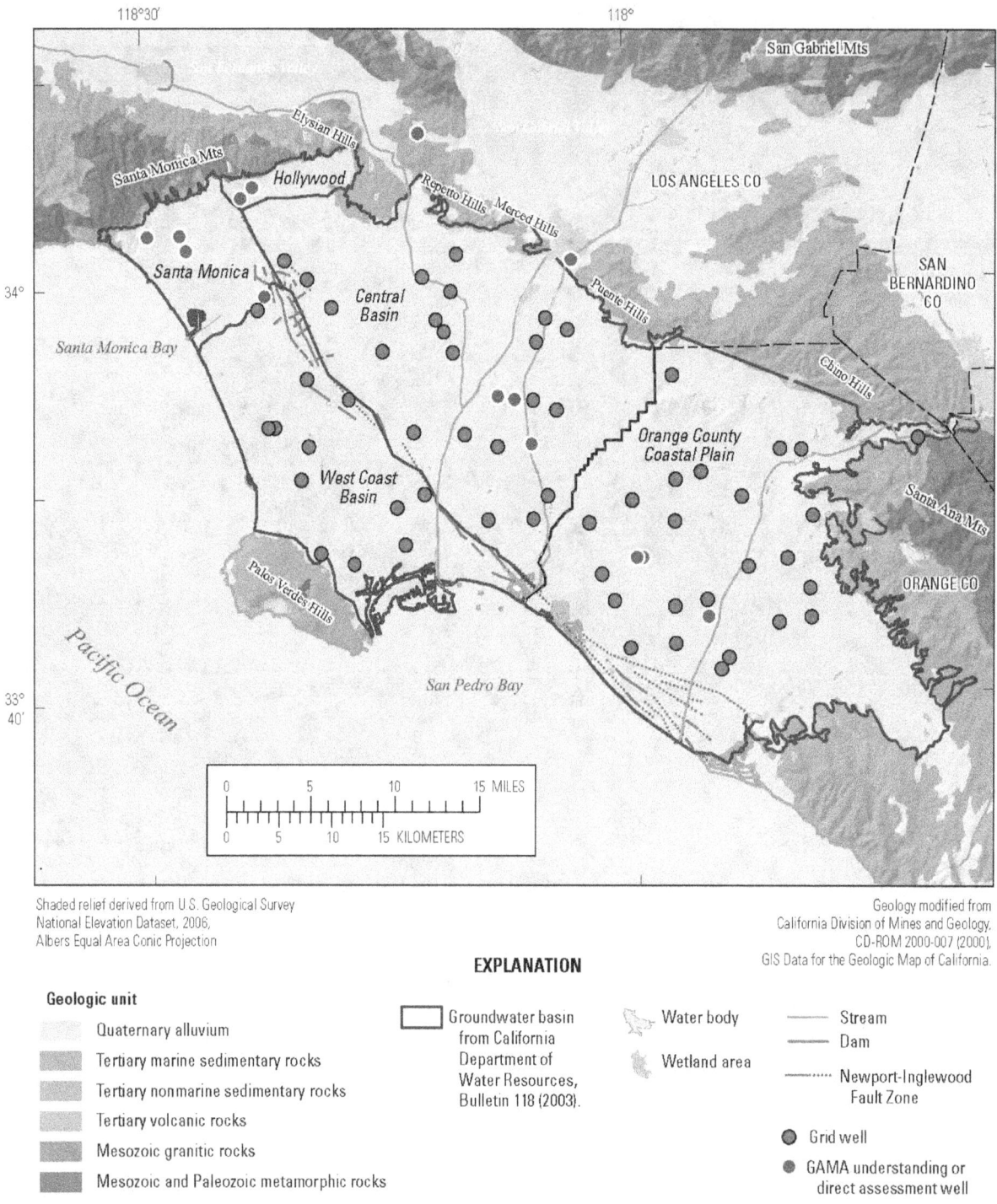

Shaded relief derived from U.S. Geological Survey
National Elevation Dataset, 2006,
Albers Equal Area Conic Projection

Geology modified from
California Division of Mines and Geology,
CD-ROM 2000-007 (2000),
GIS Data for the Geologic Map of California.

EXPLANATION

Geologic unit

Quaternary alluvium

Tertiary marine sedimentary rocks

Tertiary nonmarine sedimentary rocks

Tertiary volcanic rocks

Mesozoic granitic rocks

Mesozoic and Paleozoic metamorphic rocks

Groundwater basin
from California
Department of
Water Resources,
Bulletin 118 (2003).

Water body

Wetland area

Stream

Dam

Newport-Inglewood
Fault Zone

Grid well

GAMA understanding or
direct assessment well

Figure 3.   Coastal Los Angeles Basin study unit, 2006, California GAMA Priority Basin Project.

The climate in the CLAB study unit area is classified as Mediterranean, characterized by warm summers and cool winters. Daytime highs in the winter average about 70 degrees Fahrenheit (°F), and summer highs average between 80 and 85°F. Nearly all rainfall occurs from late autumn to early spring; virtually no precipitation falls during the summer. The average rainfall in the Coastal Los Angeles Basin area is about 15 inches (National Oceanic and Atmospheric Administration, 2007). Potential evapotranspiration in the CLAB study area exceeds precipitation on an annual basis; under natural conditions, the lower reaches of rivers that drain the basin are dry in the summer (California Department of Water Resources, 2003).

Groundwater flow is largely controlled by engineered recharge along the San Gabriel and Santa Ana Rivers and the Rio Hondo, and by groundwater pumping from the many hundreds of wells distributed across the basin. The engineered recharge sites are located on the northernmost reaches of the rivers within the boundaries of the study unit (fig. 2), and the groundwater pumping draws the water laterally and radially from these discrete sites across the study unit towards the coast (Dawson and others, 2003). Along the coast near the saltwater intrusion barriers, the direction of groundwater flow is affected by groundwater pumping and water injection in the control barriers (Shelton and others, 2001).

Methods

The *status assessment* provides a spatially unbiased assessment of groundwater quality in the primary aquifer system of the CLAB study unit. This section describes the methods used for: (1) defining groundwater quality, (2) assembling the datasets used for the *status assessment*, (3) determining which constituents warrant additional evaluation, and (4) calculating aquifer-scale proportions. Methods used for compilation of data regarding potential explanatory factors are described in appendix C.

In this study, groundwater-quality data are presented as *relative-concentrations*, the concentrations of constituents measured in groundwater relative to regulatory and non-regulatory benchmarks used to evaluate drinking-water quality. Some benchmarks are established for protection of human health, and others are established for aesthetic properties, such as taste or odor. Constituents were selected for additional evaluation in the assessment on the basis of objective criteria defined in terms of relative-concentrations. Groundwater-quality data collected by the U.S. Geological Survey for the GAMA Priority Basin Project (USGS–GAMA) and data compiled in the CDPH database are used in the *status assessment*. Two statistical approaches based on spatially unbiased equal-area grids are used to calculate aquifer-scale proportions of low, moderate, or high relative-concentrations (Belitz and others, 2010): (1) the "grid-based" approach uses one value per grid cell to represent groundwater quality, and (2) the "spatially weighted" approach uses many values per grid cell.

The CDPH database contains historical records from more than 25,000 wells, necessitating targeted retrievals to effectively access relevant water-quality data. For example, for the area representing the CLAB study unit, the historical CDPH database contains more than 502,000 records from 850 wells. The CDPH data were used in three ways in the *status assessment*: (1) to supplement the USGS data for the grid-based calculations of aquifer-scale proportions, (2) to select constituents for additional evaluation in the assessment, and (3) to provide the majority of the data used in the spatially weighted calculations of aquifer-scale proportions.

Relative-Concentrations and Water-Quality Benchmarks

Concentrations of constituents are presented as relative-concentrations in the *status assessment*:

$$Relative\ concentration = \frac{Sample\ concentration}{Benchmark\ concentration}$$

Relative-concentrations were used to provide context for the measured concentrations in the sample. Relative-concentrations less than 1 (<1.0) indicate a sample concentration less than the benchmark, and relative-concentrations greater than 1 (>1.0) indicate a sample concentration greater than the benchmark. The use of relative-concentrations also permits comparison on a single scale of constituents present at a wide range of concentrations.

Toccalino and others (2004), Toccalino and Norman (2006), and Rowe and others (2007) previously used the ratio of measured sample concentration to the benchmark concentration [either maximum contaminant levels (MCLs) or Health-Based Screening Levels (HBSLs)] and defined this ratio as the Benchmark Quotient. Relative-concentrations used in this report are equivalent to the Benchmark Quotient reported by Toccalino and others (2004) for constituents with MCLs. However, HBSLs were not used in this report because HBSLs are not currently used as benchmarks by California drinking-water regulatory agencies. Relative-concentrations can only be computed for constituents with water-quality benchmarks; therefore, constituents without water-quality benchmarks are not included in the *status assessment*.

Regulatory and non-regulatory benchmarks apply to treated water that is served to the consumer, not to untreated groundwater. However, to provide some context for the results, concentrations of constituents measured in the untreated groundwater were compared to benchmarks established by the U.S. Environmental Protection Agency (USEPA) and CDPH (U.S. Environmental Protection Agency, 1999, 2009, 2011; California Department of Public Health, 2010, 2011a). The benchmarks used for each constituent were selected in the following order of priority:

1. Regulatory, health-based CDPH and USEPA maximum contaminant levels (MCL-CA and MCL-US), action levels (AL-US), and treatment technique levels (TT-US).

2. Non-regulatory CDPH and USEPA secondary maximum contaminant levels (SMCL-CA and SMCL-US). For constituents with both recommended and upper SMCL-CA levels, the values for the upper levels were used.

3. Non-regulatory, health-based CDPH notification levels (NL-CA), USEPA lifetime health advisory levels (HAL-US), and USEPA risk-specific doses for 1:100,000 (RSD5-US).

For constituents with multiple types of benchmarks, this hierarchy may not result in selection of the benchmark with the lowest concentration. Additional information on the types of benchmarks and listings of the benchmarks for all constituents analyzed is provided by Mathany and others (2008).

For ease of discussion, relative-concentrations of constituents were classified into low, moderate, and high categories:

Category	Relative-concentrations for organic and special-interest constituents	Relative-concentrations for inorganic constituents
High	> 1	> 1
Moderate	> 0.1 and ≤ 1	> 0.5 and ≤ 1
Low	≤ 0.1	≤ 0.5

For organic and special-interest constituents, a relative-concentration of 0.1 was used as a threshold to distinguish between low and moderate relative-concentrations for consistency with other studies and reporting requirements (U.S. Environmental Protection Agency, 1998; Toccalino and others, 2004). For inorganic constituents, a relative-concentration of 0.5 was used as a threshold to distinguish between low and moderate relative-concentrations. The primary reason for using a higher threshold was to focus attention on the inorganic constituents of most immediate concern (Fram and Belitz, 2012). The

naturally occurring inorganic constituents tend to be more prevalent than organic constituents in groundwater. Although more complex classifications could be devised based on the properties and sources of individual constituents, use of a single moderate/low threshold value for each of the two major groups of constituents provided a consistent objective criteria for distinguishing constituents present at moderate rather than low concentrations.

Datasets for Status Assessment

U.S. Geological Survey Grid Wells

The primary data used for the grid-based calculations of aquifer-scale proportions of relative-concentrations were data from wells sampled by USGS-GAMA. Detailed descriptions of the methods used to identify wells for sampling are given in Mathany and others (2008). Briefly, the Central Basin, Orange County Coastal Plain, and West Coast Basin study areas each were divided into 10-mi^2 (~25-km^2) equal-area grid cells, and in each cell, one well was randomly selected for sampling to represent the cell (fig. 4) (Scott, 1990). Wells were selected to sample from the population of wells in the statewide database maintained by the CDPH. If a cell had no accessible wells listed in the CDPH database, then appropriate wells were selected from the statewide database maintained by the USGS. The CLAB study unit contained 61 grid cells, and the USGS sampled wells in 55 of those cells (USGS-grid wells). Of the 55 USGS-grid wells, 48 were listed in the CDPH database, and the other 7 wells were screened or perforated at depth intervals similar to those of wells listed in the CDPH database. USGS-grid wells were named with an alphanumeric GAMA ID consisting of a prefix identifying the study unit (CLAB), followed by a prefix identifying the study area and a number indicating the order of sample collection (fig. A1A; table A1). The following prefixes were used to identify the study areas: CB, Central Basin study area; OC, Orange County Coastal Plain study area; and WB, West Coast Basin study area. For clarity, the "CLAB" prefix is dropped from the well names on figure A1A.

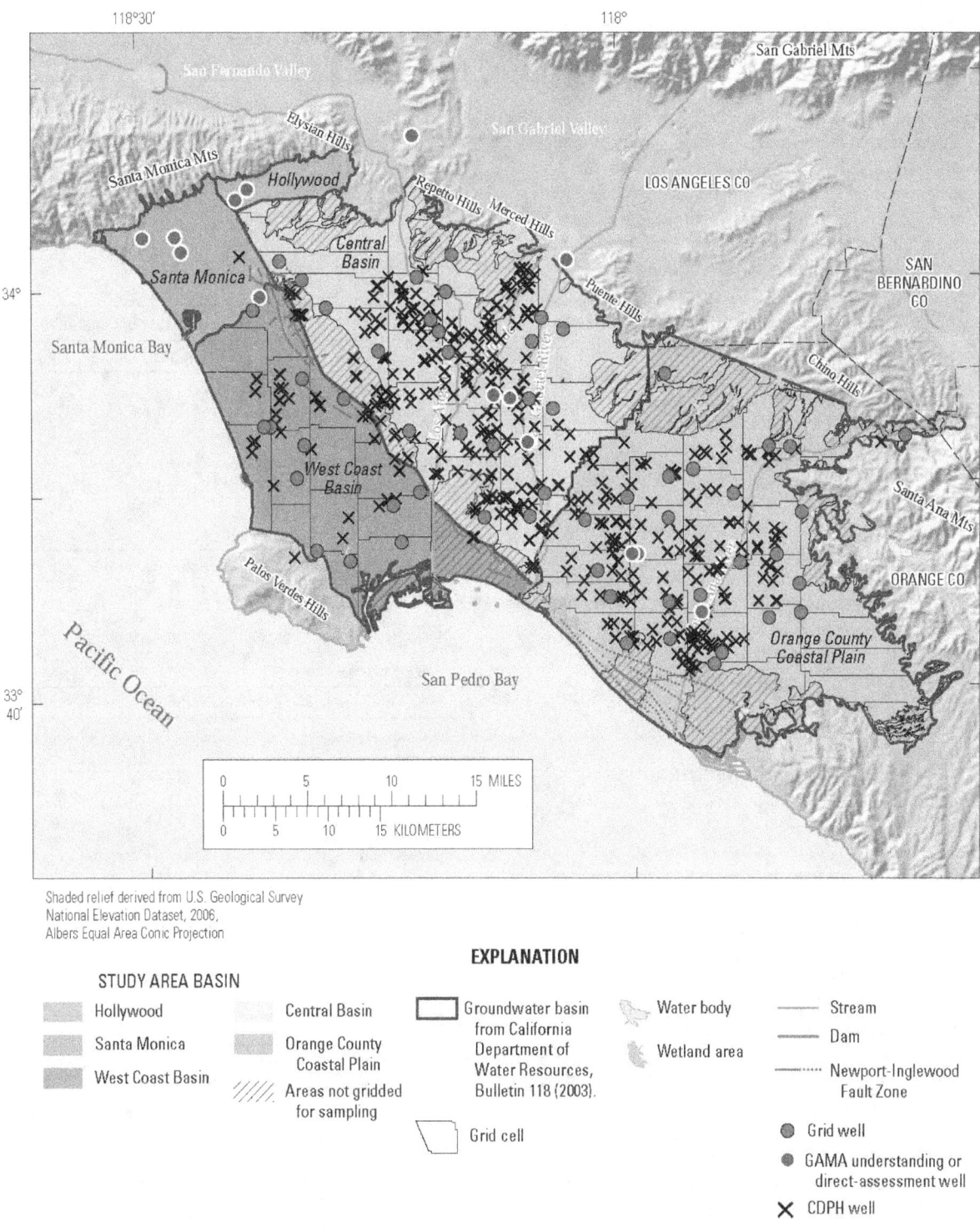

Shaded relief derived from U.S. Geological Survey
National Elevation Dataset, 2006,
Albers Equal Area Conic Projection

EXPLANATION

STUDY AREA BASIN

Hollywood

Santa Monica

West Coast Basin

Central Basin

Orange County
Coastal Plain

//// Areas not gridded
for sampling

☐ Groundwater basin
from California
Department of
Water Resources,
Bulletin 118 (2003).

⬡ Grid cell

Water body

Wetland area

——— Stream

——— Dam

·········· Newport-Inglewood
Fault Zone

● Grid well

● GAMA understanding or
direct-assessment well

✕ CDPH well

Figure 4. Locations of grid cells, California Department of Public Health (CDPH) wells, and grid, direct-assessment, and understanding wells sampled for the Coastal Los Angeles Basin study unit, 2006, California GAMA Priority Basin Project.

Six wells were sampled to assess water quality in the Hollywood Basin and Santa Monica Basin study areas where the grid-cell network approach was not implemented due to the distribution of CDPH wells and the relatively small geographic area of the basins. These wells, designated as "USGS-direct-assessment" wells, were numbered with the study unit prefix CLAB, followed by the prefix DA (for direct assessment), and a number indicating order of sample collection (fig. A1). The direct-assessment wells were not used in the aquifer-scale proportion calculations because the calculations were not applied to those two study areas.

Samples collected from USGS-grid wells were analyzed for 165 to 275 constituents (table 1). Water-quality indicators (field parameters), VOCs, pesticides, perchlorate, redox species, noble gases, and selected isotopes ("Fast" schedule on table 1) were analyzed in samples from all 55 USGS-grid wells. At five of the grid wells, additional samples were collected for analysis of gasoline oxygenates, polar pesticides, dissolved organic carbon (DOC), major and minor ions, alkalinity, trace elements, nutrients, carbon isotopes, N-nitrosodimethylamine (NDMA), 1,2,3-trichloropropane (1,2,3-TCP), and 1,4-dioxane ("Intermediate" schedule). At three of the grid wells, additional samples were collected for analysis of the constituents on the "Intermediate" schedule, plus turbidity, radioactive constituents, and microbial constituents ("Slow" schedule). The collection, analysis, and quality-control data for the analytes listed in table 1 are described by Mathany and others (2008).

California Department of Public Health Grid Wells

Data collected by USGS-GAMA at the USGS-grid wells provided part of the data used for the status assessment for inorganic constituents; the rest of the data were obtained from the CDPH database. Of the 61 grid cells, 3 cells had USGS-grid wells with the full complement of inorganic constituent data collected by USGS-GAMA, 5 cells had USGS-grid wells with USGS data for all inorganic constituents except for radioactive constituents, 47 cells had USGS-grid wells with no USGS data for inorganic constituents, and 6 cells had no USGS-grid wells. The CDPH database was queried to provide these missing data for inorganic constituents. CDPH wells with data for the most recent 3 years available at the time of sampling (June 4, 2003, through June 4, 2006) were considered. If a well had more than one analysis for a constituent in the 3-year interval, then the most recent data were selected.

The procedures used to identify suitable data from CDPH wells are described in appendix A. Briefly, the first choice was to use CDPH data from the same well as the USGS-grid well. These CDPH grid wells were labeled with the same study area prefix and number as the USGS-grid well in the cell, with the additional prefix DG (where DG refers to CDPH and USGS) (table A1; fig. A1B). If the DG well did not have all

Table 1.   Analyte groups and numbers of constituents and wells sampled for each analytical schedule, Coastal Los Angeles Basin study unit, 2006, California GAMA Priority Basin Project.

	Analytical schedule[1]		
	Fast	Intermediate	Slow
Total number of wells	50	9	10
Number of grid wells sampled	47	5	3
Number of understanding wells sampled	3	4	1
Number of direct assessment wells sampled			6

Analyte class	Number of constituents		
Inorganic constituents			
Specific conductance	1	1	1
Nutrients and dissolved organic carbon		6	6
Major ions, alkalinity, and total dissolved solids		12	12
Trace elements		25	25
Radioactive constituents[2]			5
Organic and special-interest constituents[3]			
Volatile organic compounds (VOCs)	85	88	88
Pesticides and degradates	63	116	116
Special-interest constituents	1	4	4
Geochemical and age-dating tracers			
Dissolved oxygen and temperature	2	2	2
Arsenic, chromium, and iron redox ratios	3	3	3
Tritium	1	1	1
Noble gasses (helium, neon, argon, krypton, xenon), $^3He/^4He$ of helium and tritium	7	7	7
δ^2H and $\delta^{18}O$ of water	2	2	2
pH		1	1
Carbon-14 and $\delta^{13}C$ of dissolved carbonates		2	2
Sum:	165	270	275

[1]"Fast," "intermediate," and "slow" analytical schedules refer to the amount of time required for a field crew to complete all work at a well.

[2]Both gross alpha and gross beta particle activities were measured after 72-hour and 30-day holding times; data from the 30-day measurement are used in this report. Radium activity equals the sum of the two isotopes measured: radium-226 and radium-228. Uranium activity equals the sum of the three isotopes measured: uranium-235, uranium-235, and uranium-238.

[3]The counts of organic constituents differ from those in Mathany and others (2008) because constituents analyzed by more than one analytical method are only counted once in this report. The "intermediate" and "slow" schedules included 3 gasoline oxygenates (VOCs) and 53 polar pesticides and degradates not included on the "fast" schedule. All samples also were analyzed for 14 pharmaceutical compounds and 55 potential wastewater indicator compounds; these constituents are not included in the count of constituents because results are not presented in this report. Results for pharmaceutical compounds are presented in Fram and Belitz (2011a).

the needed data, then a second well in the cell was randomly selected from the subset of CDPH wells having the most of the needed data. These CDPH grid wells were labeled with the same study area prefix and number as the USGS-grid well in the cell, with the additional prefix DPH (where DPH refers to CDPH) (table A1; fig. A1C). CDPH-grid wells in cells with no USGS-grid wells were labeled with the next number in the sequence. The combination of the USGS-grid wells and the CDPH-grid wells produced a grid-well network covering 57 of the 61 grid cells in the CLAB study unit (table A1). No accessible wells or data were available for the remaining 4 cells.

The CDPH database generally did not contain data for all missing inorganic constituents at every CDPH-grid well; therefore, the number of wells used for the grid-based assessment differed for various inorganic constituents (table 2). Although other organizations also collect water-quality data, the CDPH data is the only statewide database of groundwater-chemistry data available for comprehensive analysis.

CDPH data were not used to provide grid values for VOCs, pesticides, or special-interest constituents for the *status assessment* because a larger number of VOCs and pesticide compounds are analyzed for the USGS-GAMA Program than are available from the CDPH database (table 3). In addition, method detection limits for USGS-GAMA analyses typically were one to two orders of magnitude less than the reporting levels for analyses compiled by the CDPH (table 3). The CDPH database for the CLAB study unit contained data for 75 organic constituents that were not analyzed for by USGS-GAMA. Of these 75 constituents, 71 had no reported detections in the CDPH database.

Additional Data used for Spatially Weighted Calculations

The spatially weighted calculations of aquifer-scale proportions of relative-concentrations used data from the USGS-grid wells, from additional wells sampled by USGS-GAMA, and from all wells in the CDPH database with water-quality data during the 3-year interval June 4, 2003, through June 4, 2006. For wells with USGS and CDPH data, only the USGS data were used.

Eight non-grid wells were selected to increase sampling density in the CLAB study unit to better understand specific groundwater-quality issues. These "USGS-understanding" wells were numbered with the study unit prefix CLAB, followed by the prefix U, and a number indicating the order of sample collection (fig. A1). Two of the "DPH" CDPH-grid wells selected for the grid network also were USGS-understanding wells (table A1). The five USGS-understanding wells located in the CB or OC study areas were included in the spatially weighted aquifer-scale proportion calculations. One understanding well was located outside of the study unit and was not included in the calculations.

Table 2. Inorganic constituents and associated benchmark information, and the number of grid wells with USGS-GAMA data and CDPH data, for each constituent, Coastal Los Angeles Basin study unit, 2006, California GAMA Priority Basin Project.

[CDPH, California Department of Public Health; SMCL, Secondary Maximum Contaminant Level; USGS, U.S. Geological Survey]

Constituent	Number of cells with data from USGS GAMA[1]	Number of cells with data from CDPH database
Inorganic constituents with health-based benchmarks		
Trace elements		
Aluminum	9	44
Antimony	9	44
Arsenic	9	44
Barium	9	44
Beryllium	9	44
Boron	9	37
Cadmium	9	44
Chromium	9	40
Copper	9	44
Fluoride	9	44
Lead	9	44
Mercury	8	44
Molybdenum	10	3
Nickel	9	44
Selenium	9	44
Strontium	10	0
Thallium	9	44
Vanadium	9	37
Nutrients		
Ammonia	10	4
Nitrate[2]	8	45
Nitrite	9	43
Radioactive constituents		
Gross alpha particle activity	4	42
Gross beta particle activity	4	10
Radium activity	3	34
Radon-222 activity	4	7
Uranium[3]	10	28
Inorganic constituents with SMCL benchmarks		
Iron	9	44
Manganese	9	44
Silver	9	44
Zinc	9	44
Chloride	9	44
Specific conductance	54	2
Sulfate	9	44
Total dissolved solids (TDS)	9	44

[1]The number of cells with data from USGS-GAMA varies from 3 to 10 for different inorganic constituents because 8 grid wells were sampled by USGS-GAMA on the slow or intermediate schedules (table 1), and two of the CDPH-grid wells selected as "DPH" grid wells also were sampled by USGS-GAMA as understanding wells on the slow schedule. These "DPH" wells provided data to represent the grid cell when the "DG" CDPH-grid wells in the cell did not have data for a particular constituent.

[2]USGS-GAMA analyses were for nitrate plus nitrite; however, nitrite concentrations were negligible compared to nitrate concentrations.

[3]A conversion factor of 0.7 was used to estimate uranium activities from USGS-GAMA data for uranium concentrations.

Table 3.   Comparison of number of compounds and median reporting levels by type of constituent for data stored in the California Department of Public Health database and data collected by the Coastal Los Angeles Basin study unit, 2006, California GAMA Priority Basin Project.

[CDPH, California Department of Public Health; DLR, detection limit for the purpose of reporting; MDL, method detection level; LRL, laboratory reporting level; mg/L, milligrams per liter; µg/L, micrograms per liter]

Constituent type	CDPH		USGS GAMA		Median units
	Number of compounds	Median DLR or MDL	Number of compounds	Median LRL	
Volatile organic compounds	64	0.5	88	0.08	µg/L
Pesticides plus degradates	35	1	116	0.04	µg/L
Other organic compounds	75	1	0	None	µg/L
Special-interest compounds	4	1	4	0.2	µg/L
Nutrients	5	1	6	0.01	mg/L
Trace elements	20	8	25	0.1	µg/L

Selection of Constituents for Additional Evaluation

As many as 275 constituents were analyzed in samples from CLAB study unit wells; however, only a subset of these constituents were identified for additional evaluation in this report. Of the 275 constituents analyzed, 136 constituents did not have benchmarks (table 4). Because relative-concentrations cannot be calculated for constituents without benchmarks, these 136 constituents were not evaluated in this report. The 139 constituents having benchmarks were assessed, and a subset of these constituents were selected for additional evaluation on the basis of the following three criteria:

- Constituents present at high or moderate relative-concentrations in the CDPH database within the 3-year interval (June 4, 2003, through June 4, 2006);

- Constituents present at high or moderate relative-concentrations in the USGS-grid wells or USGS-understanding wells; or

- Organic constituents with detection frequencies of greater than 10 percent in the USGS-grid-well dataset for the study unit.

A complete list of the constituents investigated by USGS-GAMA in the CLAB study unit may be found in the CLAB Data Series Report (Mathany and others, 2008).

The CDPH database also was used to identify constituents with high relative-concentrations historically, but not currently. The historical period was defined as extending from the earliest record maintained in the CDPH database to June 4, 2003 (August 15, 1974, to June 4, 2003). Constituent concentrations may be historically high, but not currently high, because of improvement of groundwater quality with time or abandonment of wells with high concentrations. Historically high concentrations of constituents that do not otherwise meet the criteria for additional evaluation are not considered representative of potential groundwater-quality concerns in the study unit from 2003 to 2006.

For the CLAB study unit, there were 27 constituents with high concentrations reported in the CDPH database during the historical period that did not also have high concentrations reported during the current period or in the USGS-GAMA dataset (table 5). Of these 27 constituents, 17 did not meet criteria for additional evaluation in the status assessment. Many of the constituents reported at high concentrations only during the historical period were reported at high concentrations in only 1 well.

Calculation of Aquifer-Scale Proportions

The *status assessment* is intended to characterize the quality of groundwater resources in the primary aquifer system of the CLAB study unit. The primary aquifer system is defined by the depth intervals over which wells listed in the CDPH database are screened or perforated; these wells are primarily classified as municipal and community drinking-water supply wells. The use of the term "primary aquifer system" does not imply that there exists a discrete aquifer unit. In most groundwater basins, municipal and community supply wells generally are perforated at greater depths than domestic wells. However, to the extent that domestic wells are perforated over the same depth intervals as the CDPH wells, the assessments presented in this report also may be applicable to the portions of the aquifer systems used for domestic drinking-water supplies.

Two statistical approaches, grid based and spatially weighted (Belitz and others, 2010), were selected to evaluate the proportions of the primary aquifer system in the CLAB study unit with high, moderate, and low relative-concentrations of constituents relative to benchmarks. For ease of discussion, these proportions are referred to as "high, moderate, and low aquifer-scale proportions." Calculations of aquifer-scale proportions were made for individual constituents, as well as for classes of constituents. The classes consisted of groups of related individual constituents. For constituents with human-heath benchmarks, the classes included trace elements, nutrients, radioactive constituents, trihalomethanes, solvents, and herbicides.

The grid-based calculation uses the grid-well dataset assembled from the USGS-grid and CDPH-grid wells. For each constituent, the high aquifer-scale proportion was calculated by dividing the number of cells represented by a high value for that constituent by the total number of grid cells with data for that constituent. The moderate and

low aquifer-scale proportions were calculated similarly. Confidence intervals for the high aquifer-scale proportions for individual constituents were computed using the Jeffreys interval for the binomial distribution (Brown and others, 2001; Belitz and others, 2010). For calculation of high aquifer-scale proportion for a class of constituents, cells were considered high if values for any of the constituents in that class were high. Cells were considered moderate if values for any of the constituents were moderate, but no values were high. The grid-based estimate is spatially unbiased. However, the grid-based approach may not detect constituents that are present at high concentrations in small proportions of the primary aquifer system.

Table 4. Number of constituents analyzed and detected in each constituent class with each type of benchmark, Coastal Los Angeles Basin study unit, 2006, California GAMA Priority Basin Project.

[Regulatory, health-based benchmarks include: USEPA and CDPH maximum contaminant levels and USEPA action levels. Non-regulatory, health-based benchmarks include: USEPA health advisory levels and risk-specific doses at 10^{-5} and CDPH notification levels. Non-regulatory, aesthetic-based benchmarks include: USEPA and CDPH secondary maximum contaminant levels. Abbreviations: USEPA, U.S. Environmental Protection Agency; CDPH, California Department of Public Health]

	Groups of inorganic constituents							
	Nutrients		Major ions and trace elements		Radioactive constituents		Sum of inorganic constituents	
	Number of constituents							
Benchmark type	Analyzed	Detected	Analyzed	Detected	Analyzed	Detected	Analyzed	Detected
Regulatory, health-based	2	2	15	14	5	5	22	21
Non-regulatory, health-based	1	1	4	4	0	0	5	5
Non-regulatory, aesthetic-based	0	0	8	7	0	0	8	7
None	3	3	11	11	0	0	14	14
Total:	6	6	38	36	5	5	49	47

	Groups of organic constituents					
	Volatile organic compounds		Pesticides and degradates		Sum of organic constituents	
	Number of constituents					
Benchmark type	Analyzed	Detected	Analyzed	Detected	Analyzed	Detected
Regulatory, health-based	33	19	12	2	45	21
Non-regulatory, health-based	26	4	26	6	52	10
Non-regulatory, aesthetic-based	0	0	0	0	0	0
None	29	2	78	11	107	13
Total:	88	25	116	19	204	44

	Other constituents				Sum of all constituents	
	Geochemical and age-dating tracers		Special-interest constituents			
	Number of constituents					
Benchmark type	Analyzed	Detected	Analyzed	Detected	Analyzed	Detected
Regulatory, health-based	2	2	1	1	70	45
Non-regulatory, health-based	0	0	3	2	60	17
Non-regulatory, aesthetic-based	1	1	0	0	9	8
None	15	15	0	0	136	42
Total:	18	18	4	3	275	112

Table 5.   Constituents reported at concentrations greater than benchmarks historically (August 15, 1974, to June 4, 2003) in the CDPH database, Coastal Los Angeles Basin study unit, 2006, California GAMA Priority Basin Project.

[High value, concentration above benchmark. Benchmarks: MCL-US, USEPA maximum contaminant level; MCL-CA, CDPH maximum contaminant level; SMCL-CA, CDPH secondary maximum contaminant level; NL-CA, CDPH notification level; HAL-US, USEPA lifetime health advisory level; RSD5-US, USEPA risk-specific dose at 10^{-5}. Abbreviations: CDPH, California Department of Public Health; USEPA, U.S. Environmental Protection Agency; mg/L, milligrams per liter; μg/L, micrograms per liter]

Constituent	Benchmark			Date of most recent high value	Number of wells with historical data	Number of wells with a high value
	Type	Value	Units			
Inorganic constituents						
Ammonia[1]	HAL-US	24.7	mg/L	06-06-1995	128	1
Antimony[2]	MCL-US	6	μg/L	12-22-1992	661	1
Barium[3]	MCL-US	1,000	μg/L	11-15-1987	680	1
Beryllium	MCL-US	4	μg/L	02-16-1994	660	1
Boron[4]	NL-CA	1,000	μg/L	02-27-1997	600	1
Cadmium	MCL-US	5	μg/L	11-08-2002	680	12
Chromium	MCL-CA	50	μg/L	12-04-2001	680	5
Fluoride[5]	MCL-CA	2	mg/L	09-04-2002	684	8
Lead	AL-US	15	μg/L	11-08-2002	680	17
Selenium	MCL-US	50	μg/L	03-30-1989	680	2
Sulfate[6]	SMCL-CA	500	mg/L	02-28-1997	683	1
Thallium	MCL-US	2	μg/L	08-24-1998	660	2
Organic constituents						
Atrazine	MCL-CA	1	μg/L	02-10-1993	684	19
Benzene	MCL-CA	1	μg/L	11-08-1999	694	2
Chloroform	MCL-US	80	μg/L	09-27-2000	694	2
Cyanazine[7,8]	HAL-US	1	μg/L	10-15-1985	113	1
Cyanide[8]	MCL-CA	150	μg/L	12-26-2001	630	2
Dibromoethane (EDB)	MCL-US	0.05	μg/L	05-20-2003	675	1
cis-1,2-Dichloroethene	MCL-CA	6	μg/L	03-07-2002	687	1
Dichloromethane	MCL-US	5	μg/L	09-27-2000	694	3
Dieldrin[7]	RSD5-US	0.05	μg/L	10-15-1985	510	1
Dinoseb[7]	MCL-US	7	μg/L	10-15-1985	552	1
Endrin[8]	MCL-US	2	μg/L	06-16-1990	653	1
Lindane[8]	MCL-US	0.2	μg/L	09-22-1988	651	1
Methyl tert-butyl ether	MCL-CA	13	μg/L	10-28-1998	661	1
Simazine[9]	MCL-US	4	μg/L	04-14-1988	684	1
Total trihalomethanes	MCL-US	80	μg/L	09-27-2000	601	4

[1] High value for ammonia was 300; the same well had a value of 0.36 six months earlier and 0 six months later. The CDPH database does not include a field for units; thus, results reported in micrograms per liter and results reported in milligrams per liter may be reported together.

[2] High value was 200; the same well had a non-detection (< 6) six months later.

[3] High value for barium was 280,280, which likely is a typographical error. The next highest value is 680.

[4] High value for boron was 2,060; the same well had value of 0 three years earlier and 110 two years later.

[5] Four wells had high values for fluoride between 2 and 10. Four wells had high values between 110 and 600; all other values for these four wells were low (well A, 600 and eight values between 0 and 0.36; well B, 300, 0.30, and 0.34; well C, 200 and five values between 0.20 and 0.32; and well D, 110 and nine values between 0.31 and 0.50). The four values between 110 and 600 may be in nanograms per liter rather than in micrograms per liter. The CDPH database does not contain a field specifying the units of measurement, and some constituents may be reported in different units by different laboratories.

[6] High value for sulfate was 1,400; the same well had a value of 61 forty months earlier and a value of 430 forty-two months later.

[7] High values for cyanazine (150), dieldrin (0.05), and dinoseb (100) were reported in the same well on the same day. No other detections of these three constituents were reported in the database. The reported values were equal to the reporting limits for other samples collected during the same time period. The values for cyanzine and dinoseb may be in nanograms per liter rather than micrograms per liter.

[8] Cyanazine, cyanide, endrin, and lindane were not analyzed by USGS-GAMA in this study unit.

[9] High value for simazine was 3,002, which likely is a typographical error. The next highest value is 1.6.

The spatially weighted calculation uses the dataset assembled from all CDPH and USGS-GAMA wells. For each constituent, the high aquifer-scale proportion was calculated by computing the proportion of wells with high values in each cell and then averaging the proportions for all cells (Isaaks and Srivastava, 1989; Belitz and others, 2010). The moderate aquifer-scale proportion was calculated similarly. Confidence intervals for spatially weighted detection frequencies of high concentrations are not described in this report. For calculation of high aquifer-scale proportion for a class of constituents, values for wells were considered high if the values for any of the constituents in that class were high. Values for wells were considered moderate if the values for any of the constituents were moderate, but no values for wells were high.

In addition, for each constituent, the raw detection frequencies of high and moderate values for individual constituents were calculated using the same dataset as used for the spatially weighted calculations. However, raw detection frequencies are not spatially unbiased because the wells in the CDPH database are not uniformly distributed throughout the CLAB study unit (fig. 4). For example, if a constituent were present at high concentrations in a small region of the aquifer with a high density of wells, the raw detection frequency of high values would be greater than the high aquifer-scale proportion. Raw detection frequencies are provided for reference but were not used to assess aquifer-scale proportions (see appendix B for additional details about the statistical approaches).

The grid-based high aquifer-scale proportions were used to represent proportions in the primary aquifer system unless the spatially weighted proportions were significantly different than the grid-based values. Significantly different results were defined as follows:

- If the grid-based high aquifer-scale proportion was zero and the spatially weighted proportion was non-zero, then the spatially weighted result was used. This situation can happen when the concentration of a constituent is high in a small fraction of the primary aquifer system.

- If the grid-based high aquifer-scale proportion was non-zero and the spatially weighted proportion was outside the 90-percent confidence interval (based on the Jeffreys interval for the binomial distribution), then the spatially weighted proportion was used.

The grid-based moderate and low proportions were used in most cases because the reporting levels for many organic constituents and some inorganic constituents in the CDPH database were higher than the threshold between moderate and low categories. However, if the grid-based moderate proportion was zero and the spatially weighted proportion non-zero, then the spatially weighted value was used as a minimum estimate for the moderate proportion.

Potential Explanatory Factors

Data for a finite set of potential explanatory factors were compiled: land use, well depth, depth to top of screened or perforated interval, density of septic systems, density of formerly leaking underground fuel tanks, groundwater age, oxidation-reduction condition, and pH. Methods used for assigning values of potential explanatory factors to the CLAB study unit wells are described in appendix C. Statistical assessments of the correlations among potential explanatory factors and between potential explanatory factors and water quality are not presented in this report.

Land Use

Land use was described by three land-use types: urban, agricultural, and natural (appendix C). Percentages of the three types were calculated for the study unit and study areas and for areas within a radius of 500 meters (m; 500-m buffers) around wells (Johnson and Belitz, 2009).

Land use in the CLAB study unit is 86 percent urban, 12.7 percent natural, and 1.6 percent agricultural (figs. 5, 6). Most of the agricultural land use is located in the southeastern portion of the OC study area. Small areas of natural land use are scattered throughout the study unit. The largest contiguous areas of natural land use are the Seal Beach Naval Weapons Station, Anaheim Bay National Wildlife Refuge, and Bolsa Chica Ecological Reserve in the coastal portion of the OC study area.

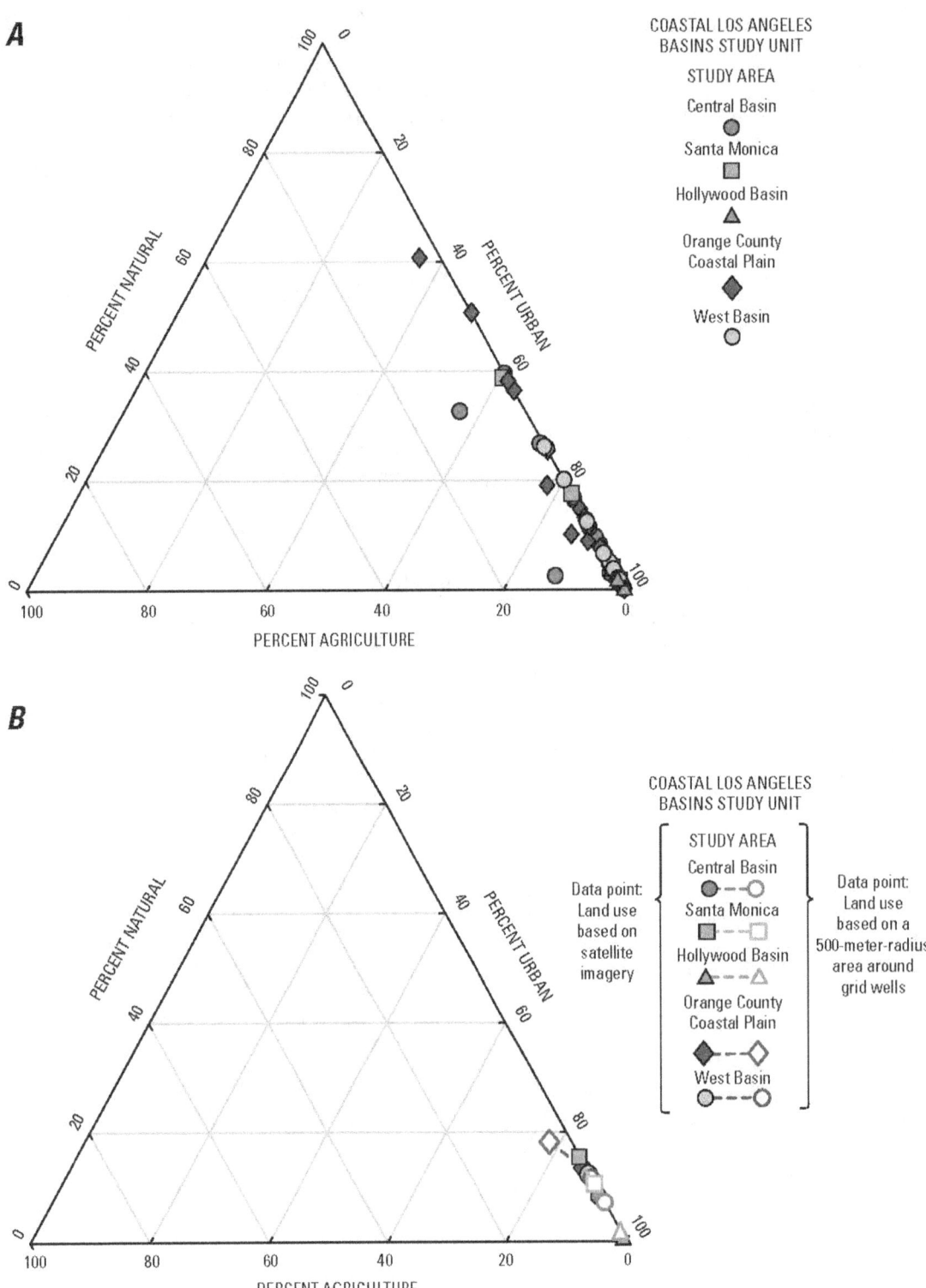

Figure 5. Proportions of urban, agricultural, and natural land-use categories for (*A*) wells in the Coastal Los Angeles Basin study unit and (*B*) in study unit areas and a 500-meter radius around wells.

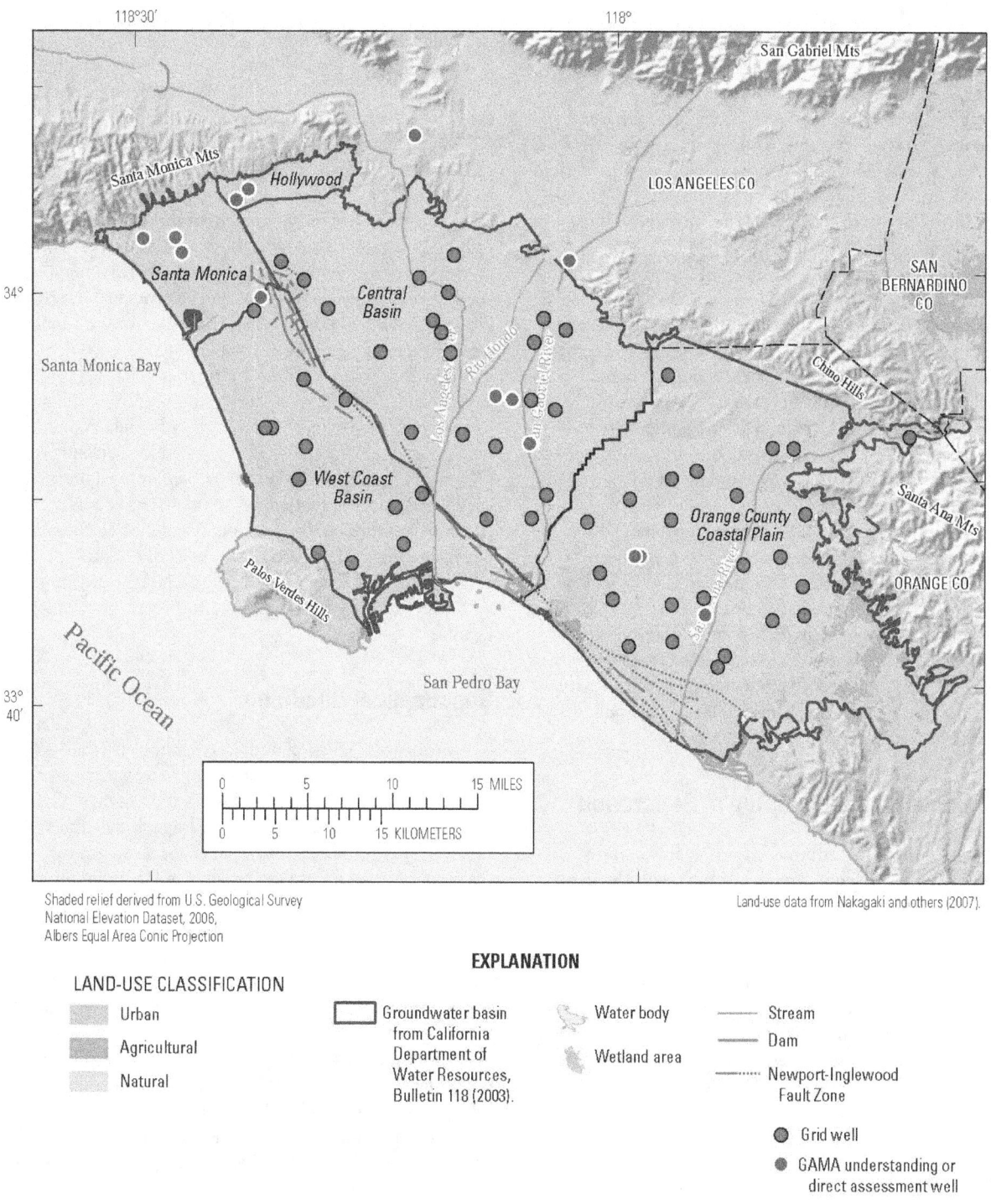

Shaded relief derived from U.S. Geological Survey
National Elevation Dataset, 2006,
Albers Equal Area Conic Projection

Land-use data from Nakagaki and others (2007).

EXPLANATION

LAND-USE CLASSIFICATION

Urban

Agricultural

Natural

Groundwater basin
from California
Department of
Water Resources,
Bulletin 118 (2003).

Water body

Wetland area

Stream

Dam

Newport-Inglewood
Fault Zone

Grid well

GAMA understanding or
direct assessment well

Figure 6.   Land-use map of the Coastal Los Angeles Basin study unit and locations of wells.

For the CB and WB study areas, the percentage of urban land use in the study area as a whole, 93 percent and 88 percent, respectively, was the same as the average percentage of urban land use in the areas within a 500-m radius of the grid wells. For the OC study area, the area around the grid wells averaged 85 percent urban, whereas the study area as a whole was 78 percent urban. Land use around individual grid wells ranged from 36 percent to 100 percent urban, and 42 of the 55 grid wells were surrounded by greater than 85 percent urban land use (fig. 5; table C1).

The density of leaking or formerly leaking underground fuel tanks and the density of septic tanks in the 500-m radius area around a well may be indicators of potential sources of anthropogenic contaminants at the land surface. The density of leaking or formerly leaking underground fuel tanks around grid wells ranged from 0.1 tanks per square kilometer (tanks/km²) to 6.2 tanks/km², and the median density was 1.6 tanks/km² (table C1). The density of septic tanks around grid wells ranged from 0.0 to 62 tanks/km², and the median density was 0.2 tanks/km² (table C1).

Dawson and others (2003) demonstrated that groundwater from wells closer to the sites of engineered recharge on the San Gabriel River, the Rio Hondo, and the Santa Ana River was more influenced by recharged water than groundwater from wells farther away. Distances from engineered recharge sites were not tabulated for this report because the sites are linear features parallel to the primary direction of groundwater flow, thus different zero points for measured flow paths may be selected.

Well Depth and Depth to Top of Perforation

Well construction information was available for 64 of the 69 wells sampled in the CLAB study unit. Depth of USGS-grid wells ranged from 98 to 1,680 feet (ft; 30 to 512 m) below land surface; the median depth was 822 ft (251 m) (fig. 7, table C1). Depths to the tops of the perforations ranged from 60 to 684 ft (18 to 208 m), with a median of 397 ft (121 m). The perforation length was as much as 1,075 ft (328 m), with a median of 420 ft (128 m). The understanding and direct-assessment wells generally had shallower depths and shallower depths to the top of the screened or perforated interval than the grid wells.

Groundwater Age

Groundwater samples were assigned age classifications on the basis of the tritium and carbon-14 contents of the samples (see section "Groundwater Age Classification" in appendix C). Groundwater with tritium activity greater than 1 tritium unit (TU) was defined as "modern," and groundwater with tritium activity less than 1 TU was defined as "pre-modern." Modern groundwater contains a substantial component of water recharged since 1952. The presence of pre-modern groundwater also was identified using the carbon-14 data: samples with percentage of modern carbon less than 90 percent were considered to contain a substantial component of pre-modern groundwater. Samples with tritium activity greater than 1 TU and modern carbon percentage less than 90 percent would be classified as "mixed."

Samples from 31 wells were classified as pre-modern groundwater. Samples from 9 wells were classified as modern-age groundwater, and samples from the remaining 29 wells were classified as modern or mixed because carbon-14 data were not available to assess the presence or absence of pre-modern groundwater. There were no significant differences in well depth or depth to the top of the screened interval between the 31 wells with pre-modern groundwater and the 38 wells with modern or mixed groundwater (Wilcoxon rank-sum test, p>0.05; fig. 8A). In the CB and OC study areas, wells with modern or mixed groundwater generally were located further inland than wells with pre-modern groundwater (fig. 8B). Dawson and others (2003) and Hamlin and others (2005) interpreted this pattern as reflecting flow of modern water recharged in engineered recharge facilities on the Rio Hondo and the San Gabriel and Santa Ana Rivers near the western edge of the study unit. This modern recharge water has not yet reached the coastal side of the study areas because they are furthest from the recharge facilities.

Geochemical Condition

Oxidation-reduction (redox) conditions for the 69 wells sampled by USGS-GAMA were classified by using an abridged version of the redox classification framework of McMahon and Chapelle (2008) and Jurgens and others (2009) (table C2). The 33 wells with dissolved oxygen (DO) concentration less than or equal to 0.5 milligram per liter (mg/L) were classified as anoxic, and the 36 wells with DO greater than 0.5 mg/L were classified as oxic. Wells classified as anoxic were further classified by type of anoxic redox process occurring (suboxic, nitrate-reducing, manganese-reducing, or iron-reducing) if sufficient data for inorganic constituents were available. The range of geochemical conditions may reflect natural spatial variability or mixing of water from different depths in wells that have long perforated intervals. Most of the wells in the OC study area had oxic groundwater, and most of the wells in the WB study area had anoxic groundwater (fig. 9A). Redox conditions were not classified for wells from the CDPH database because no DO data were available.

The pH ranged from 6.5 to 8.8 in the USGS-grid wells and USGS-direct-assessment wells, and from 7.0 to 9.1 in CDPH-database wells (fig. 9B, table C2). The pH value of water indicates the acidity or basicity of the water. Values of pH greater than 8.0 primarily occurred in groundwater from the OC study area and the southern corner of the Central Basin study area (fig. 9). Values of pH less than 7.0 were rarely observed.

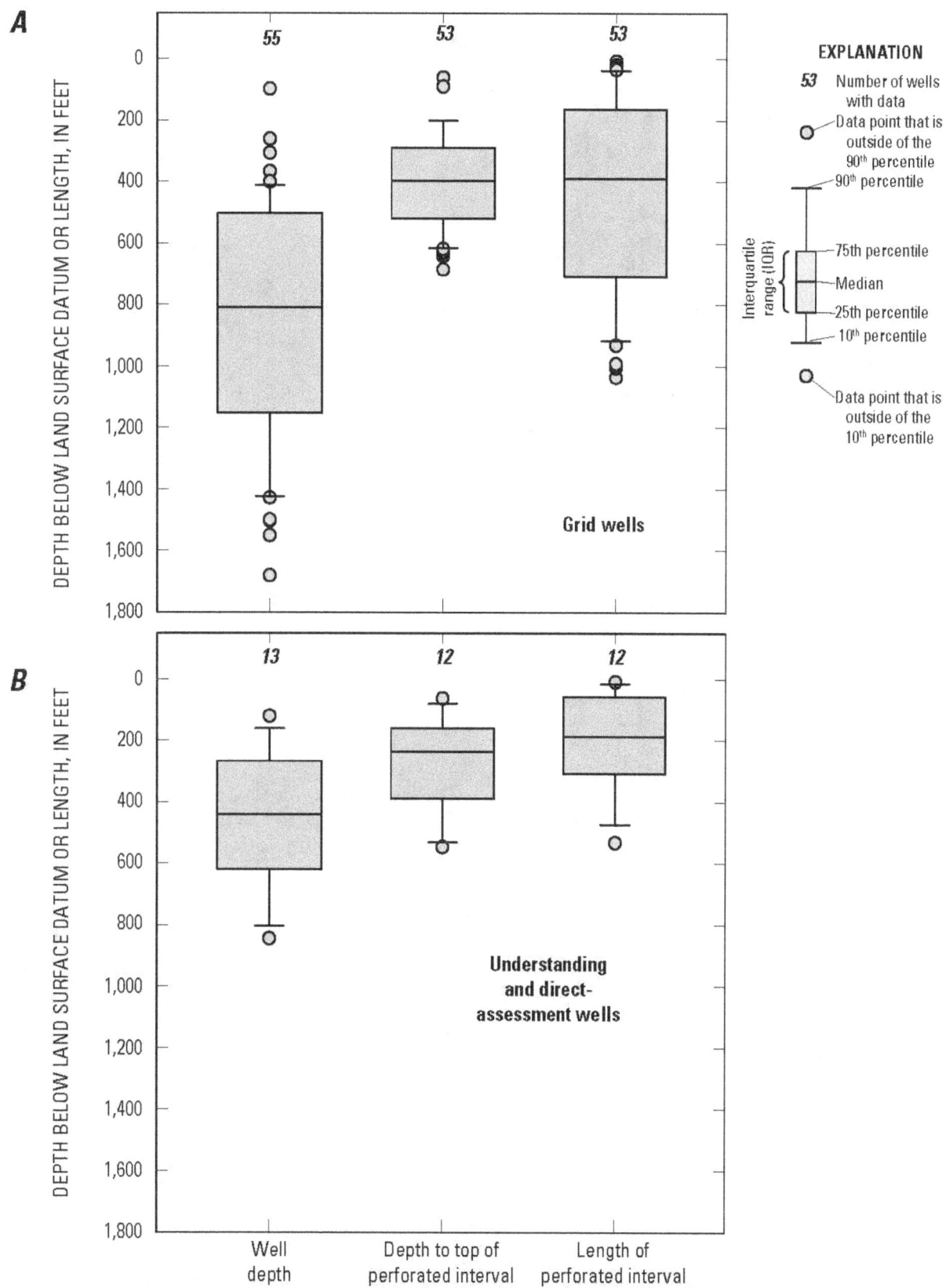

Figure 7. Construction characteristics for (*A*) grid, and (*B*) direct-assessment and understanding wells, Coastal Los Angeles Basin study unit, 2006, California GAMA Priority Basin Project.

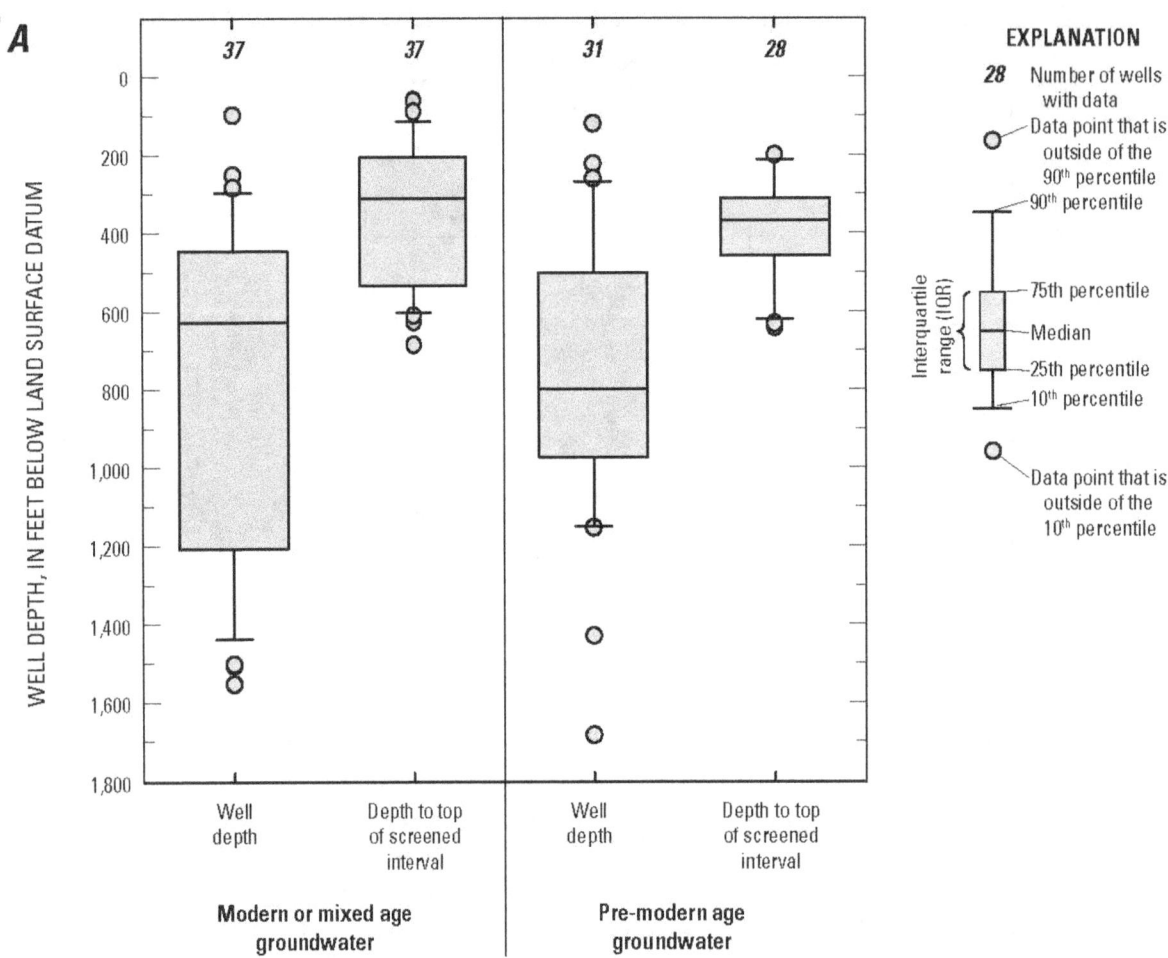

Figure 8.   Relation of classified groundwater age to (A) well depth and depth to top of perforations, and (B) wells with classified age distributions in two age categories, Coastal Los Angeles Basin study unit, 2006, California GAMA Priority Basin Project.

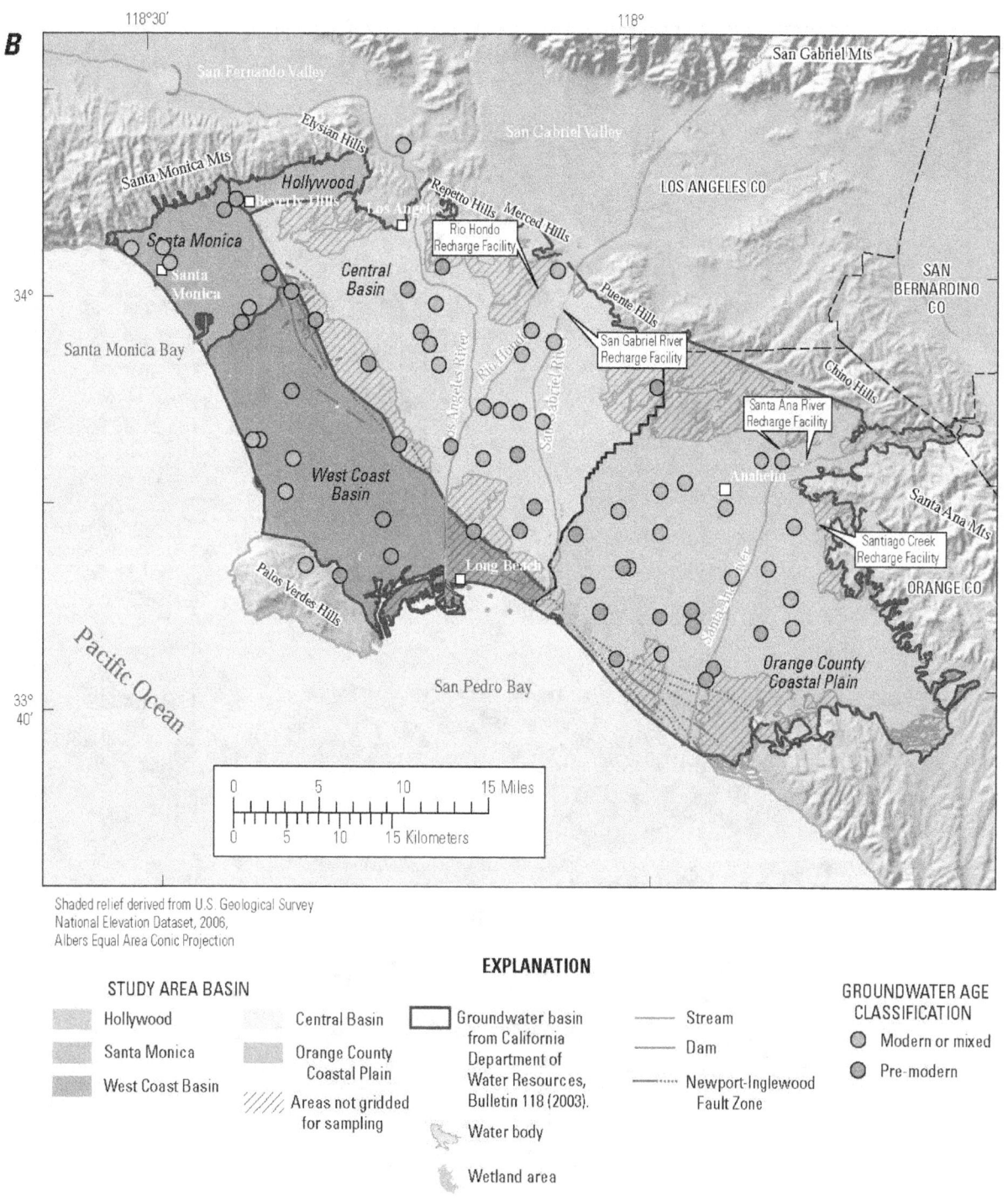

Shaded relief derived from U.S. Geological Survey
National Elevation Dataset, 2006,
Albers Equal Area Conic Projection

EXPLANATION

STUDY AREA BASIN

Hollywood

Santa Monica

West Coast Basin

Central Basin

Orange County
Coastal Plain

Areas not gridded
for sampling

Groundwater basin
from California
Department of
Water Resources,
Bulletin 118 (2003).

Water body

Wetland area

Stream

Dam

Newport-Inglewood
Fault Zone

GROUNDWATER AGE
CLASSIFICATION

Modern or mixed

Pre-modern

Figure 8.—Continued

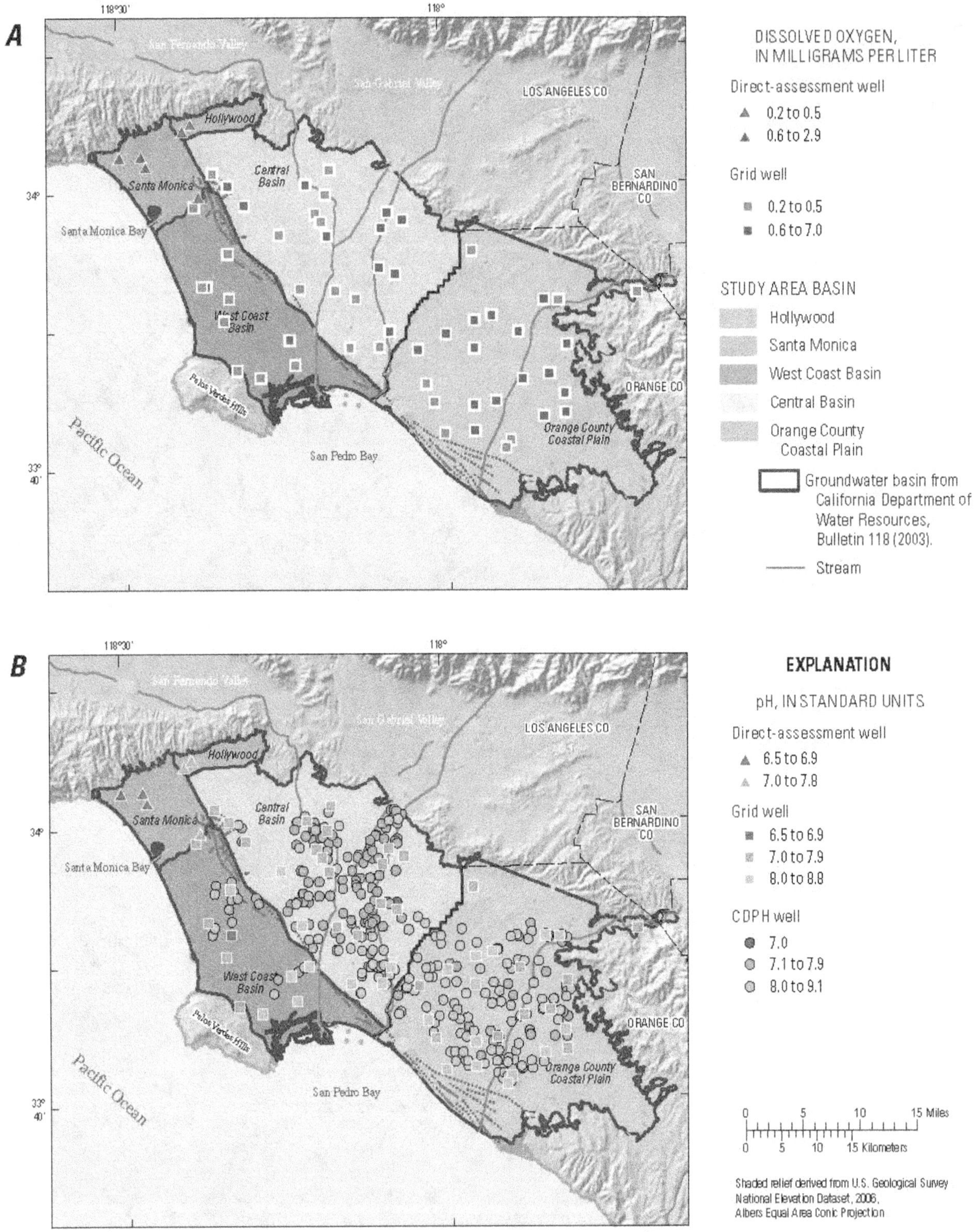

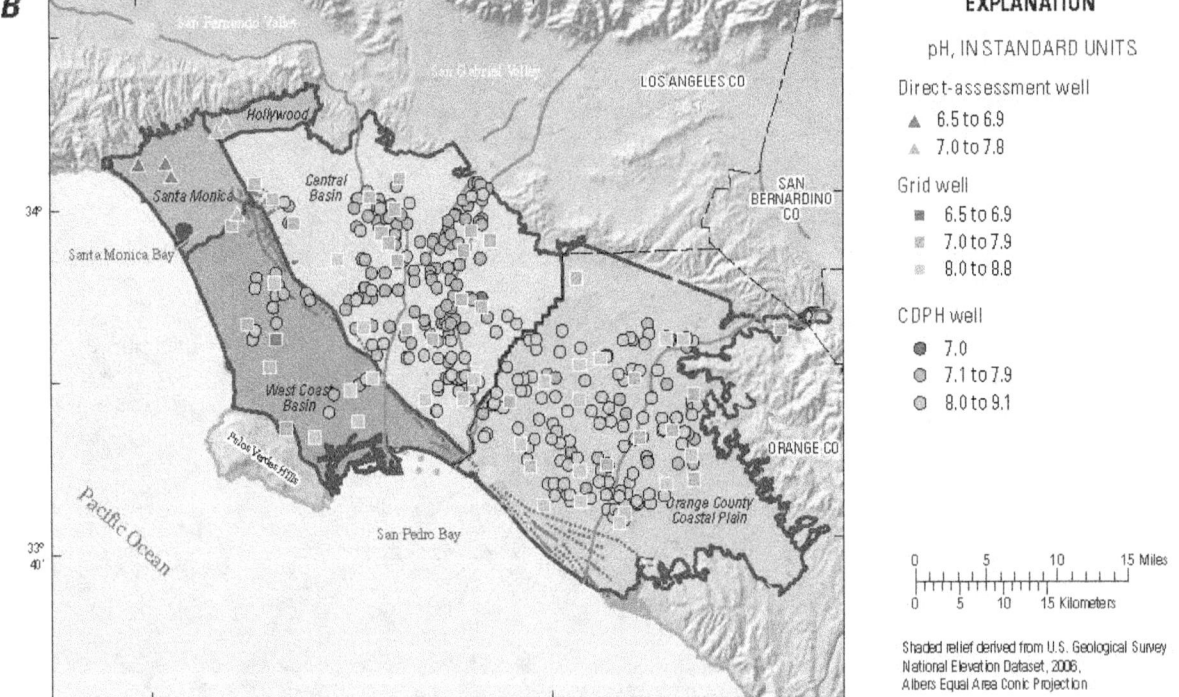

Figure 9. Values of (*A*) dissolved oxygen and (*B*) pH in grid wells, understanding wells, direct-assessment wells, and other wells in the CDPH database, Coastal Los Angeles Basin study unit, 2006, California GAMA Priority Basin Project.

Status of Water Quality

The *status assessment* was designed to identify the constituents or classes of constituents most likely to be of water-quality concern because of their high relative-concentrations or their prevalence. USGS sample analyses, plus additional data from the CDPH database were included in the assessment of groundwater quality for the CLAB study unit. The spatially distributed, randomized approach to grid-well selection and data analysis yields a view of groundwater quality in which all areas of the primary aquifer system are weighted equally; regions with a high density of groundwater use or with high density of potential contaminants were not preferentially sampled (Belitz and others, 2010). The summary of detection data from the Santa Monica and Hollywood study areas is included, but *status assessment* methods were not applied because of insufficient well coverage in grid cells.

The following discussion of the *status assessment* results is divided into results for inorganic and organic constituents. The assessment begins with a survey of how many constituents were detected at any concentration compared to the number analyzed and a graphical summary of the relative-concentrations of constituents detected in the grid wells. Results are presented for the subset of constituents that met criteria for selection for additional evaluation based on relative-concentration, or for organic constituents, prevalence. Results for the direct assessment of the Hollywood and Santa Monica study areas are then presented.

The aquifer-scale proportions calculated by using the spatially weighted approach were within the 90-percent confidence intervals for their respective grid-based aquifer high proportions for all the constituents listed in table 6, providing evidence that the grid-based and spatially weighted approaches yield statistically equivalent results.

Inorganic Constituents

Inorganic constituents generally occur naturally in groundwater, although their concentrations may be influenced by human activities as well as natural factors. Forty-seven of the 49 inorganic constituents analyzed by the USGS-GAMA were detected in the CLAB study unit. Of these 47 detected constituents, 26 had regulatory or non-regulatory health-based benchmarks, 7 had non-regulatory aesthetic-based benchmarks, and 14 had no established benchmarks (table 4). Most of the constituents without benchmarks are major or minor ions that are present in nearly all groundwater.

Eleven inorganic constituents were selected for additional evaluation in the *status assessment* because they were detected at moderate or high concentrations in the grid wells: the trace elements arsenic and boron, the nutrient nitrate, the radioactive constituents uranium and gross alpha particle activity, and the constituents with aesthetic-based benchmarks (SMCLs), iron, manganese, sulfate, specific conductance, chloride, and total dissolved solids (table 6, figs. 10, 11). An additional six inorganic constituents were selected for additional evaluation because they were reported at high or moderate concentrations in the CDPH database during the period June 4, 2003, to June 4, 2006: aluminum, fluoride, lead, mercury, nickel, and vanadium (table 6). Inorganic constituents having human-health benchmarks, as a group (trace elements, radioactive constituents, and nutrients), had high relative-concentrations in 5.6 percent of the primary aquifer system, moderate relative-concentrations in 26 percent, and low relative-concentrations in 68 percent (table 7). Inorganic constituents having aesthetic-based benchmarks, as a group, had high relative-concentrations in 18 percent of the primary aquifer system, moderate relative-concentrations in 47 percent, and low relative-concentrations in 35 percent. The spatial distributions of concentrations of selected inorganic constituents are shown in figures 12A–E.

Table 6. Aquifer-scale proportions calculated by using grid-based and spatially weighted methods for constituents detected at high relative-concentrations during the most recent 3 years of available data (June 4, 2003–June 4, 2006) from the California Department of Public Health (CDPH) database, detected at high or moderate relative-concentrations in grid wells, or organic constituents detected in greater than 10 percent of USGS-grid wells sampled, Coastal Los Angeles Basin study unit, 2006, California GAMA Priority Basin Project.

[Grid-based aquifer proportions for organic constituents are based on samples collected by the U.S. Geological Survey from 55 grid wells during June–November 2006. Spatially weighted aquifer proportions are based on CDPH data from the period June 1, 2003, to June 1, 2006, in combination with grid-well and understanding-well data. High; concentrations greater than benchmark: moderate, concentrations less than benchmark and greater than 0.1 (for organic constituents) or 0.5 (for inorganic constituents) of benchmark; low, concentrations less than 0.1 (for organic constituents) or 0.5 (for inorganic constituents) of benchmark; %, percent; MCL-US; U.S. Environmental Protection Agency maximum contaminant level; MCL-CA, CDPH maximum contaminant level; AL-US, U.S. Environmental Protection Agency (USEPA) action level; HAL-US, USEPA Lifetime Health Advisory; SMCL-CA, CDPH secondary maximum contaminant level; mg/L, milligrams per liter; µg/L, micrograms per liter; pCi/L, picocuries per liter; µS/cm, microsiemens per centimeter; NDMA, N-Nitrosodimethylamine]

Constituent	Benchmark type	Benchmark value	Units	Raw detection frequency[1]			Spatially weighted aquifer proportions[1]			Grid-based aquifer proportions			90% confidence interval for grid-based high proportion[2]	
				Number of wells	Percent moderate	Percent high	Number of cells	Proportion moderate	Proportion high	Number of cells	Proportion moderate	Proportion high	Lower limit	Upper limit
Inorganic constituents														
Trace elements														
Aluminum	MCL-CA	1,000	µg/L	366	0	[3]0.3	53	0	[3]0.3	53	0	0	0	2.5
Arsenic	MCL-US	10	µg/L	378	6.1	2.7	52	5.2	3.6	53	3.8	1.9	0.3	7.1
Boron	NL-CA	1,000	µg/L	273	0.7	0	47	0.5	0	46	2.2	0	0	2.9
Fluoride	MCL-CA	2	mg/L	368	0.5	0	53	0.2	0	53	0	0	0	2.5
Lead	AL-US	15	µg/L	372	0.3	0	53	0.2	0	53	0	0	0	2.5
Mercury	MCL-US	2	µg/L	364	0	0.3	53	0	0.5	52	0	0	0	2.6
Nickel	MCL-CA	100	µg/L	372	0	[4]0.3	53	0	[4]0.1	53	0	0	0	2.5
Vanadium	NL-CA	50	µg/L	278	0	[5]0.4	48	0	[5]0.3	46	0	0	0	2.9
Nutrients														
Nitrate	MCL-US	10	mg/L	484	5.2	0.6	53	4.8	0.8	53	1.9	1.9	0.3	7.1
Uranium and radioactive constituents														
Uranium	MCL-CA	20	pCi/L	214	14	0.9	44	9.9	1.2	38	13	0	0	3.5
Gross alpha activity[6]	MCL-US	15	pCi/L	310	20	1.6	47	22.0	2.3	46	20	2.2	0.4	8.2
Adjusted gross alpha activity[6]	MCL-US	15	pCi/L	310	0.3	0	47	0.4	0	46	2.2	0	0	2.9
Inorganic constituents with SMCL benchmarks														
Iron	SMCL-CA	300	µg/L	384	6.3	4.2	53	6.2	9.0	53	5.7	9.4	4.4	18
Manganese	SMCL-CA	50	µg/L	388	12	12	53	14	17	53	15	15	8.4	25
Sulfate	SMCL-CA	500	mg/L	369	0.8	0	53	1.7	0	53	1.9	0	0	2.5
Specific conductance	SMCL-CA	1,600	µS/cm	360	35	0.5	57	44	1.2	56	45	1.8	0.3	6.8
Chloride	SMCL-CA	500	mg/L	369	0.5	0.3	53	2.5	0.6	53	1.9	1.9	0.3	7.1
Total dissolved solids (TDS)	SMCL-CA	1,000	mg/L	373	33	0.5	53	38	1.3	53	38	1.9	0.3	7.1

Table 6. Aquifer-scale proportions calculated by using grid-based and spatially weighted methods for constituents detected at high relative-concentrations during the most recent 3 years of available data (June 4, 2003–June 4, 2006) from the California Department of Public Health (CDPH) database, detected at high or moderate relative-concentrations in grid wells, or organic constituents detected in greater than 10 percent of USGS-grid wells sampled, Coastal Los Angeles Basin study unit, 2006, California GAMA Priority Basin Project.—Continued

[Grid-based aquifer proportions for organic constituents are based on samples collected by the U.S. Geological Survey from 55 grid wells during June–November 2006. Spatially weighted aquifer proportions are based on CDPH data from the period June 1, 2003, to June 1, 2006, in combination with grid-well and understanding-well data. High; concentrations greater than benchmark; moderate, concentrations less than benchmark and greater than 0.1 (for organic constituents) or 0.5 (for inorganic constituents) of benchmark; low, concentrations less than 0.1 (for organic constituents) or 0.5 (for inorganic constituents) of benchmark; %, percent; MCL-US: U.S. Environmental Protection Agency maximum contaminant level; MCL-CA, CDPH maximum contaminant level; AL-US, U.S. Environmental Protection Agency (USEPA) action level; HAL-US, USEPA Lifetime Health Advisory; SMCL-CA, CDPH secondary maximum contaminant level; mg/L, milligrams per liter; µg/L, micrograms per liter; pCi/L, picocuries per liter µS/cm, microsiemens per centimeter; NDMA, *N*-Nitrosodimethylamine]

Constituent	Benchmark type	Benchmark value	Units	Raw detection frequency [1]			Spatially weighted aquifer proportions [1]			Grid-based aquifer proportions			90% confidence interval for grid-based high proportion [2]	
				Number of wells	Percent moderate	Percent high	Number of cells	Proportion moderate	Proportion high	Number of cells	Proportion moderate	Proportion high	Lower limit	Upper limit
Organic and special-interest constituents														
Solvents														
1,1-Dichloroethane	MCL-CA	5	µg/L	495	0.6	0	57	0.5	0	55	0	0	0	2.4
1,1-Dichloroethene	MCL-CA	6	µg/L	495	1.6	0.6	57	1.8	0.5	55	1.8	0	0	2.4
1,2-Dichloroethane	MCL-CA	0.5	µg/L	495	0.2	0.6	57	0.1	0.3	55	0	0	0	2.4
cis-1,2-Dichloroethene	MCL-CA	6	µg/L	495	0.6	0	57	0.3	0	55	0	0	0	2.4
Perchloroethene (PCE)	MCL-US	5	µg/L	498	9.6	1.2	57	6.3	1.1	55	5.5	0	0	2.4
Carbon tetrachloride	MCL-CA	0.5	µg/L	495	0	0.6	57	0	1.0	55	0	1.8	0.3	6.9
Trichloroethene (TCE)	MCL-US	5	µg/L	495	8.7	1.8	57	6.9	1.7	55	13	0	0	2.4
Vinyl chloride	MCL-CA	0.5	µg/L	495	0.2	0	57	0.1	0	55	0	0	0	2.4
Trihalomethanes														
Chloroform	MCL-US	80	µg/L	495	0	0	57	0	0	55	0	0	0	2.4
Bromodichloro-methane	MCL-US	80	µg/L	495	0.2	0	57	0.4	0	55	0	0	0	2.4
Total trihalomethanes	MCL-US	80	µg/L	495	1.0	0	57	2.2	0	55	0	0	0	2.4
Fuel Components														
Methyl *tert*-butyl ether (MTBE)	MCL-CA	13	µg/L	500	0.2	0	57	0.4	0	55	0	0	0	2.4
Herbicides														
Atrazine	MCL-CA	1	µg/L	381	1.0	0	56	1.2	0	54	5.5	0	0	2.5
Prometon	HAL-US	100	µg/L	236	0	0	54	0	0	54	0	0	0	2.5
Simazine	MCL-US	4	µg/L	381	0	0	56	0	0	54	0	0	0	2.5
Tebuthiuron	HAL-US	500	µg/L	61	0	0	54	0	0	54	0	0	0	2.5

Table 6. Aquifer-scale proportions calculated by using grid-based and spatially weighted methods for constituents detected at high relative-concentrations during the most recent 3 years of available data (June 4, 2003–June 4, 2006) from the California Department of Public Health (CDPH) database, detected at high or moderate relative-concentrations in grid wells, or organic constituents detected in greater than 10 percent of USGS-grid wells sampled, Coastal Los Angeles Basin study unit, 2006, California GAMA Priority Basin Project.—Continued

[Grid-based aquifer proportions for organic constituents are based on samples collected by the U.S. Geological Survey from 55 grid wells during June–November 2006. Spatially weighted aquifer proportions are based on CDPH data from the period June 1, 2003, to June 1, 2006, in combination with grid-well and understanding-well data. High; concentrations greater than benchmark; moderate, concentrations less than benchmark and greater than 0.1 (for organic constituents) or 0.5 (for inorganic constituents) of benchmark; low, concentrations less than 0.1 (for organic constituents) or 0.5 (for inorganic constituents) of benchmark; %, percent; MCL-US; U.S. Environmental Protection Agency maximum contaminant level; MCL-CA, CDPH maximum contaminant level; AL-US, U.S. Environmental Protection Agency (USEPA) action level; HAL-US, USEPA Lifetime Health Advisory; SMCL-CA, CDPH secondary maximum contaminant level; mg/L, milligrams per liter; µg/L, micrograms per liter; pCi/L, picocuries per liter µS/cm, microsiemens per centimeter; NDMA, *N*-Nitrosodimethylamine]

Constituent	Benchmark Type	Benchmark Value	Units	Raw detection frequency [1]			Spatially weighted aquifer proportions [1]			Grid-based aquifer proportions			90% confidence interval for grid-based high proportion [2]	
				Number of wells	Percent moderate	Percent high	Number of cells	Proportion moderate	Proportion high	Number of cells	Proportion moderate	Proportion high	Lower limit	Upper limit
Special-interest constituents														
1,4-Dioxane[7]	NL-CA	1	µg/L	38	2.6	50	16	1.2	32	7	0	14	2.6	44
Perchlorate	MCL-CA	6	µg/L	357	6.7	0.6	57	12	0.5	55	35	0	0	2.4
N-Nitroso-dimethylamine[7]	NL-CA	0.01	µg/L	20	20	0	15	6.7	0	7	14	0	0.0	17

[1] Based on the most recent data for each CDPH well during the period June 4, 2003, to June 4, 2006, combined with GAMA grid and understanding well data.

[2] Based on the Jeffreys interval for the binomial distribution (Brown and others, 2001).

[3] One well had a reported detection of aluminum with concentration 1,100 µg/L; the same well had a non-detection (<5 µg/L) reported six weeks later.

[4] One well had a reported detection of nickel with concentration 250 µg/L; the other six samples for the same well had reported concentrations <10 µg/L or were non-detections <10 µg/L.

[5] One well had a reported detection of vanadium with concentration 66 µg/L; the same well had a detection with concentration 16 µg/L six months earlier.

[6] The MCL-US for gross alpha activity applies to adjusted gross alpha particle activity, which is equal to measured gross alpha activity minus uranium activity. Results for adjusted and unadjusted gross alpha activity are presented here. Most other Scientific Investigations Reports for GAMA Priority Basin Project study units presented results only for unadjusted gross alpha activity.

[7] The aquifer-scale proportions for 1,4-dioxane and NDMA may not be representative because few cells had wells with data for these constituents.

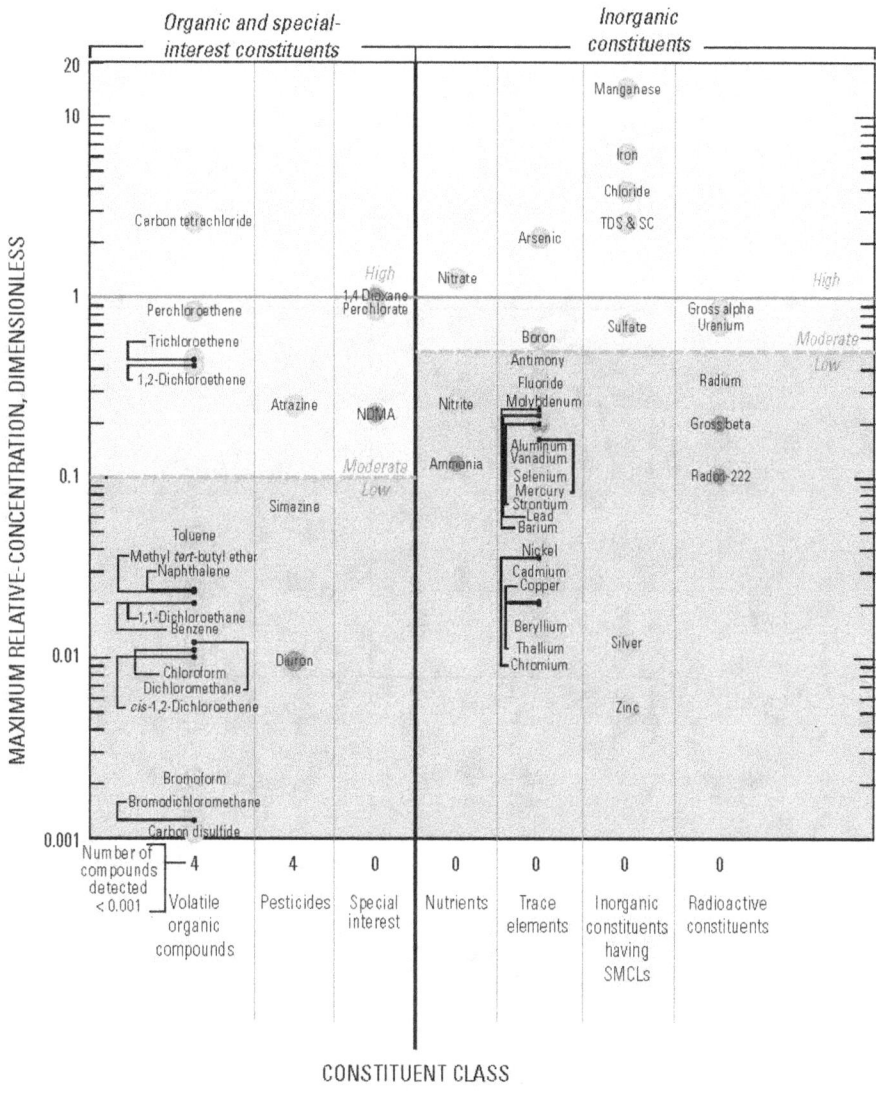

Figure 10. Maximum relative-concentration in grid wells for constituents detected, by type of constituent, Coastal Los Angeles Basin study unit, 2006, California GAMA Priority Basin Project.

EXPLANATION

Silver — **Constituent with analyses in >37 grid wells**—Name and center of symbol is location of data unless indicated by following location line:

Radon-222 — **Constituent with analyses in <15 grid wells and wells are not spatially representative**—Name and center of symbol is location of data

SMCL, secondary maximum contaminant level; upper water-quality benchmark used for calculating relative-concentrations for constituents with an upper and recommended SMCL; >, greater than; <, less than; TDS, total dissolved solids; SC, specific conductance

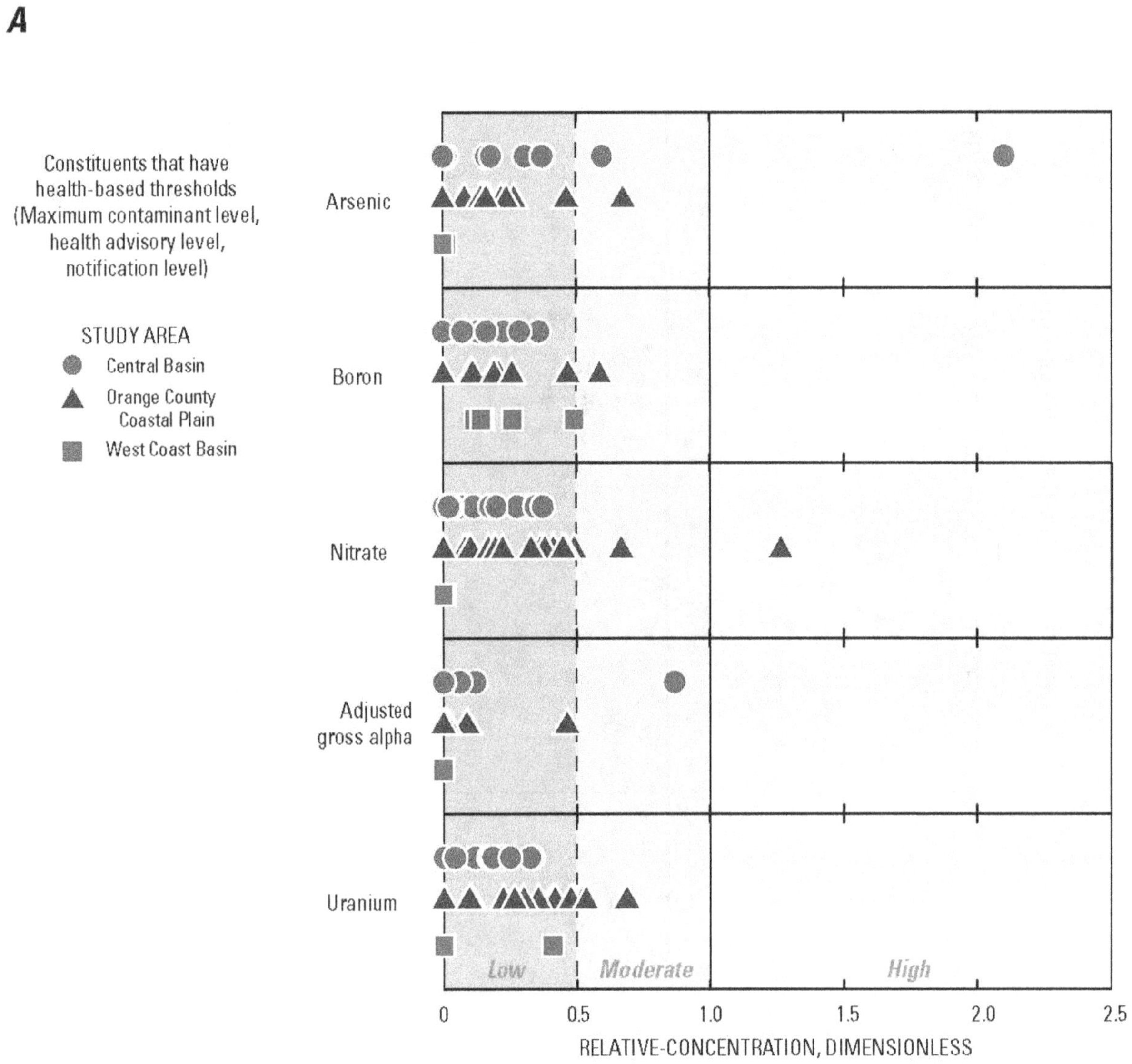

A

Constituents that have
health-based thresholds
(Maximum contaminant level,
health advisory level,
notification level)

STUDY AREA
● Central Basin
▲ Orange County
  Coastal Plain
■ West Coast Basin

Arsenic

Boron

Nitrate

Adjusted
gross alpha

Uranium

Low   *Moderate*   *High*

0   0.5   1.0   1.5   2.0   2.5

RELATIVE-CONCENTRATION, DIMENSIONLESS

Figure 11.   Relative-concentrations in grid wells of (*A*) selected trace elements, radioactive constituents, and nutrients, with human-health benchmarks and (*B*) selected constituents with aesthetic-based benchmarks, Coastal Los Angeles Basin study unit, 2006, California GAMA Priority Basin Project.

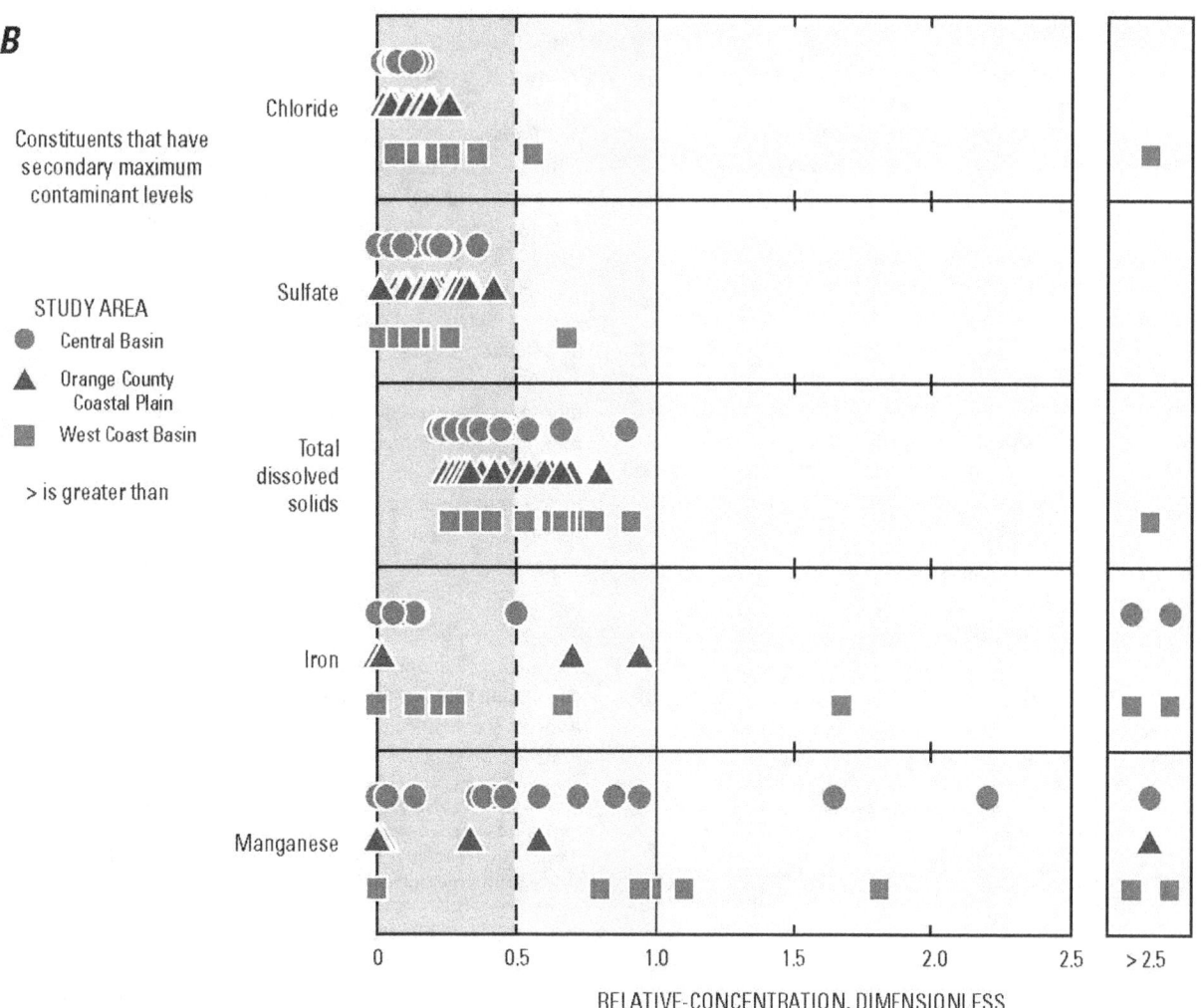

Figure 11.—Continued

Trace Elements

Trace elements, as a class, had high relative-concentrations (for one or more constituents) in 1.9 percent of the primary aquifer system, moderate values in 5.6 percent, and low values in 93 percent (table 7). Arsenic was the only trace element present at high relative-concentrations in the grid-well network.

Arsenic is a semi-metallic trace element. Natural sources of arsenic in groundwater include dissolution of arsenic-bearing minerals and desorption of arsenic from mineral surfaces. Pyrite, the most common sulfide mineral in aquifer materials, may have arsenic concentrations as high as several weight percent. Potential anthropogenic sources of arsenic include copper ore smelting, coal combustion, arsenical pesticides, arsenical veterinary pharmaceuticals, and wood preservatives (Welch and others, 2000). An estimated 8 percent of groundwater resources used for drinking water in the United States have high relative-concentrations of arsenic [>10 micrograms per liter (µg/L)] (Focazio and others, 1999). Arsenic was detected at high relative-concentrations in 1.9 percent of the primary aquifer system and at moderate relative-concentrations in 3.8 percent (table 6). Most of the wells with high or moderate relative-concentrations of arsenic were located in the southern part of the CB study area (fig. 12A).

Boron was not detected at high relative-concentrations and was detected at moderate relative-concentrations in 2.2 percent of the primary aquifer system (table 6). The trace elements selected for additional evaluation in the status assessment on the basis of high or moderate relative-concentrations reported in the CDPH database each had high aquifer-scale proportions (aluminum, mercury, nickel, and vanadium) or moderate aquifer-scale proportions (fluoride and lead) of less than 1 percent.

Table 7A. Summary of aquifer-scale proportions for inorganic constituent classes, Coastal Los Angeles Basin study unit, 2006, California GAMA Priority Basin Project.

[Relative-concentration categories: high, concentration of at least one constituent in a class greater than water-quality benchmark value; moderate, concentration of at least one constituent in class greater than half of water-quality benchmark and no constituents in class with concentration greater than benchmark; low, concentrations of all constituents in class less than or equal to half of benchmark, including non-detections. Abbreviations: SMCL, secondary maximum contaminant level]

Constituent class	Number of cells	Aquifer-scale proportion (percent)		
		Low values	Moderate values	High values
Inorganic constituents with health-based benchmarks				
Trace elements	54	93	5.6	1.9
Nutrients	53	96	1.9	1.9
Uranium and radioactive constituents[1]	49	78	20	2.0
Any constituent[2]	54	68	26	5.6
Inorganic constituents with SMCL benchmarks				
Salinity indicators[3]	57	51	47	1.8
Trace metals[4]	53	66	15	19
Any constituent	57	35	47	18

[1] Results for uranium and radioactive constituents as a class use unadjusted gross alpha activity for consistency with Scientific Investigations Reports and Fact Sheets for other GAMA Priority Basin Project study units. If adjusted gross alpha activity were used, the aquifer-scale proportions for radioactive constituents as a class would be 1.1 percent high, 11 percent moderate, and 88 percent low.

[2] Results for inorganic constituents with health-based benchmarks as a class use unadjusted gross alpha activity for consistency with Scientific Investigations Reports and Fact Sheets for other GAMA Priority Basin Project study units. If adjusted gross alpha activity were used, the aquifer-scale proportions for inorganic constituents as a class would be 3.7 percent high, 19 percent moderate, and 77 percent low.

[3] Total dissolved solids, specific conductance, sulfate, and chloride.

[4] Manganese, iron, zinc, and silver.

Table 7B. Summary of aquifer-scale proportions for organic constituent classes with health-based benchmarks, Coastal Los Angeles Basin study unit, 2006, California GAMA Priority Basin Project.

[Relative-concentration categories: high, concentration of at least one constituent in a class greater than water-quality benchmark value; moderate, concentration of at least one constituent in class greater than one-tenth of water-quality benchmark and no constituents in a class with concentration greater than benchmark; low, concentrations of all constituents in a class less than or equal to one-tenth of benchmark, including non-detections]

Constituent class	Number of cells	Aquifer-scale proportion (percent)			
		Not detected	Low values	Moderate values	High values
Trihalomethanes	55	55	43	[1]2.2	0
Solvents	55	58	27	11	[1]3.7
Gasoline components	55	80	20	[1]0.4	0
Any volatile organic compound	55	42	44	11	[1]3.7
Herbicides	54	60	35	5.5	0
Any organic constituent	55	40	44	13	[1]3.7

[1] Spatially weighted value.

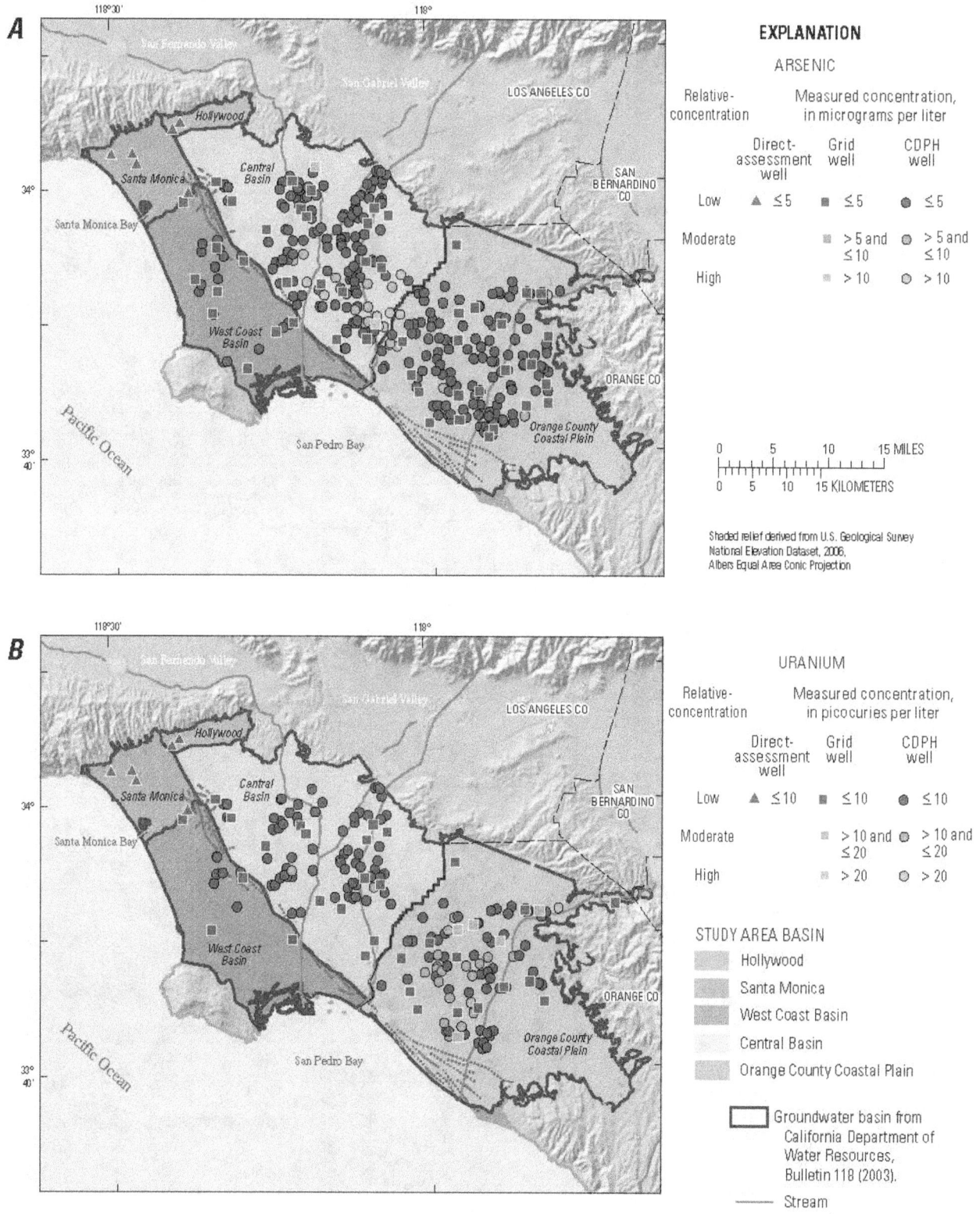

Figure 12. Concentration values of selected inorganic constituents in grid wells and direct-assessment wells, and the most recent analysis, June to November 2006, for CDPH wells, Coastal Los Angeles Basin study unit, 2006, California GAMA Priority Basin Project: (*A*) arsenic, (*B*) uranium, (*C*) nitrate, (*D*) total dissolved solids, and (*E*) manganese.

Figure 12.—Continued

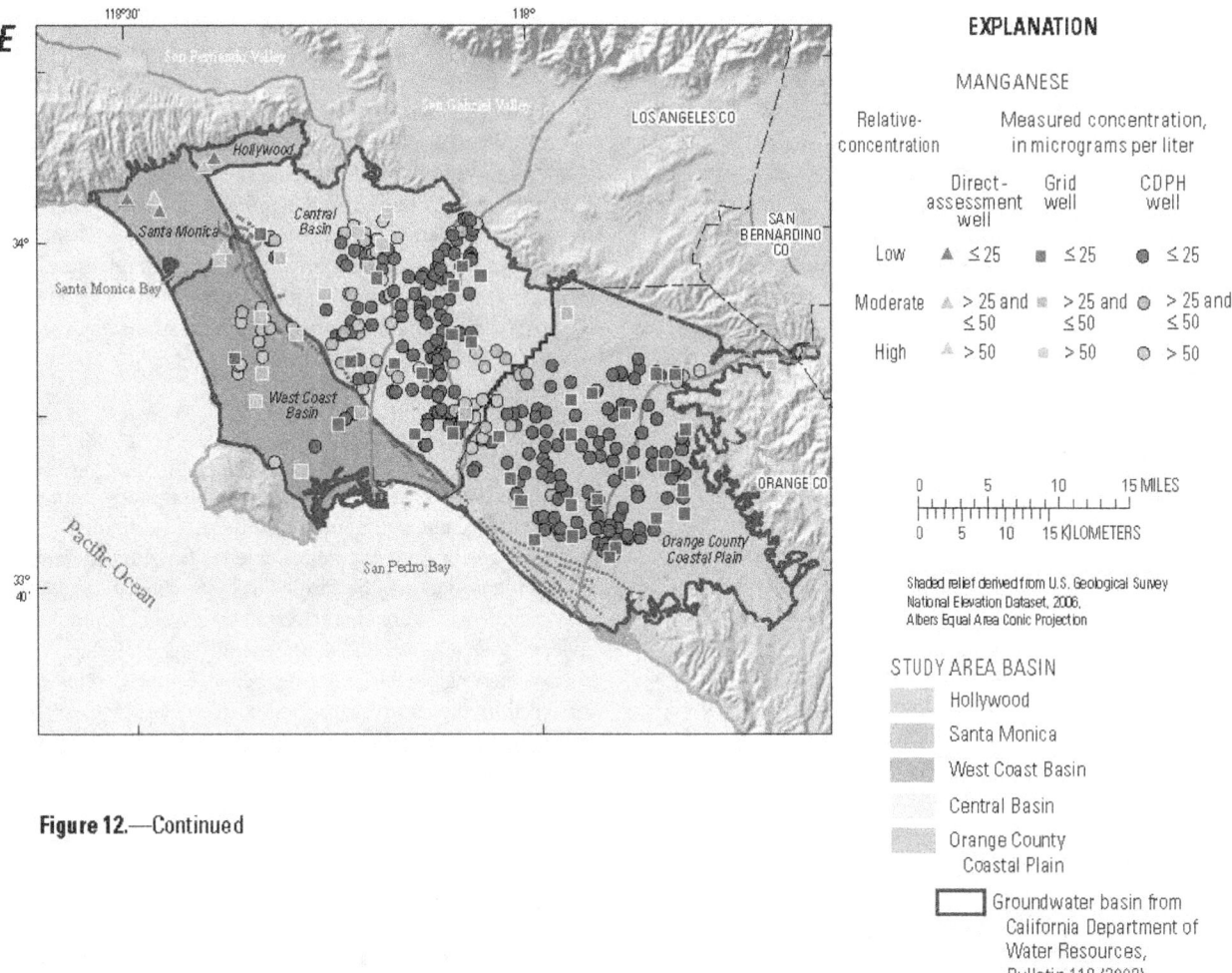

EXPLANATION

MANGANESE

Relative-concentration

Measured concentration, in micrograms per liter

	Direct-assessment well	Grid well	CDPH well
Low	▲ ≤ 25	■ ≤ 25	● ≤ 25
Moderate	▲ > 25 and ≤ 50	■ > 25 and ≤ 50	● > 25 and ≤ 50
High	▲ > 50	■ > 50	● > 50

Shaded relief derived from U.S. Geological Survey
National Elevation Dataset, 2006.
Albers Equal Area Conic Projection

STUDY AREA BASIN

- Hollywood
- Santa Monica
- West Coast Basin
- Central Basin
- Orange County Coastal Plain
- Groundwater basin from California Department of Water Resources, Bulletin 118 (2003).
- —— Stream

Figure 12.—Continued

Uranium and Radioactive Constituents

Uranium and radioactive constituents, as a class, had high relative-concentrations (for one or more constituents) in 2.0 percent of the primary aquifer system, moderate values in 20 percent, and low values in 78 percent (table 7).

The MCL-US [15 picocuries per liter (pCi/L)] for gross alpha particle activity applies to adjusted gross alpha particle activity, which is equal to the measured gross alpha particle activity minus uranium activity (U.S. Environmental Protection Agency, 2009b). Data collected by USGS-GAMA and data compiled in the CDPH database are reported as gross alpha particle activity without correction for uranium activity. Gross alpha is used a screening tool to determine whether other radioactive constituents must be analyzed. For regulatory purposes, analysis of uranium is only required if gross alpha particle activity is greater than 15 pCi/L; therefore, the CDPH database contains far more data for gross alpha particle activity than for uranium. As a result, it is not always possible to calculate adjusted gross alpha particle activity. For this reason, gross alpha data without correction for uranium are the primary data used in the status assessments made by USGS-GAMA for Priority Basin Project study units.

USGS-GAMA reports data for gross alpha particle activity counted 72 hours or 30 days after sample collection. Regulatory sampling for gross alpha particle activity permits use of quarterly composite samples (U.S. Environmental Protection Agency, 2009b); thus, the USGS-GAMA gross alpha 30-day count data may be more appropriate to use when combining USGS-GAMA and CDPH datasets. Gross alpha particle activity in a groundwater sample may change with time after sample collection due to radioactive decay and ingrowth (activity may increase or decrease depending on sample composition and holding time) (Arndt, 2010).

Uranium and gross alpha particle activity were the radioactive constituents present at high relative-concentrations. Natural sources of uranium to groundwater include dissolution of uranium-bearing minerals and desorption of uranium from mineral surfaces. Anthropogenic activities may increase uranium concentrations in groundwater by changing the chemistry of water recharging the aquifer (Jurgens and others, 2010). Uranium was detected at high relative-concentrations (spatially weighted) in 1.2 percent of the primary aquifer system and at moderate relative-concentrations in 13 percent. Nearly all of the wells with high or moderate relative-concentrations of uranium were located in the OC study area (fig. 12B).

Gross alpha particle activity was detected at high relative-concentrations in 2.2 percent of the primary aquifer system and at moderate relative-concentrations in 20 percent (table 6). The aquifer-scale proportions for adjusted gross alpha particle activity were lower: no high values were reported, and the moderate aquifer-scale proportion was 2.2 percent. The large difference between high and moderate aquifer-scale proportions between unadjusted and adjusted gross alpha particle activity suggests that most of the alpha particle activity in the samples was from uranium.

Nutrients

Nutrients as a class were detected at high relative-concentrations in 1.9 percent of the primary aquifer system and at moderate relative-concentrations in 1.9 percent (table 7). The only nutrient detected at high or moderate relative-concentrations was nitrate. Most of the wells with high or moderate relative-concentrations of nitrate were located in the OC study area (fig. 12C). Nitrate has both natural and anthropogenic sources to groundwater; however, concentrations greater than 2 mg/L (relative-concentration of 0.2) generally are considered to indicate presence of nitrate from anthropogenic sources (Mueller and Helsel, 1996). Potential anthropogenic sources of nitrate include use of fertilizers in agricultural and urban areas, nitrate in water used for engineered recharge, seepage from septic and sewage systems, and animal and human wastes.

Constituents with SMCL Benchmarks

The major ions chloride and sulfate and TDS and specific conductance have recommended and upper SMCL-CA values. In this report, data were compared to the upper values. TDS was present at high relative-concentrations in 1.9 percent of the primary aquifer system and at moderate relative-concentrations in 38 percent (table 6). The grid well with a high relative-concentration of TDS also had a high relative-concentration of chloride. Two wells with a moderate relative-concentration of TDS also had a moderate relative-concentration of chloride or sulfate.

In the WB study area, wells with high and moderate relative-concentrations of TDS were located on the seaward side of the study area (fig. 12D). Reichard and others (2003) observed the same pattern and concluded that there may be multiple sources of the salts, including seawater intrusion, water from fine-grained marine sediments, dissolution of evaporate minerals, and recharge of evaporated irrigation water. In the OC study area, wells with moderate relative-concentrations of TDS primarily were located on the inland and central portions of the study area (fig. 12D). Hamlin and others (2002) suggested the concentrations of TDS in this portion of the basin may reflect the composition of the imported surface water and water from the Santa Ana River used in the engineered recharge facilities.

Manganese was present at high relative-concentrations in 15 percent of the primary aquifer system and at moderate relative-concentrations in 15 percent (table 6). The high and moderate aquifer-scale proportions for iron were 9.4 and 5.7 percent, respectively. Most wells with high relative-concentrations of manganese were located in the WB study area, along the boundary between the CB and OC study areas, or along the Los Angeles River in the CB study area (fig. 12E). Anoxic conditions may result in release of manganese (and iron) from aquifer materials into groundwater. The areas with high relative-concentrations of manganese also were areas with anoxic groundwater (fig. 9A).

Organic Constituents

The organic constituents assessed in this study primarily are volatile organic compounds (VOCs) and pesticides. VOCs may be present in paints, solvents, fuels, refrigerants, and fumigants and may be formed as byproducts of water disinfection. VOCs are characterized by a volatile nature, or tendency to evaporate, and they generally persist longer in groundwater than in surface water because groundwater is isolated from the atmosphere. Pesticides are used to control weeds, fungi, or insects in agricultural and urban settings. One or more organic constituents were found in 26 of the 55 grid wells (47 percent) sampled in the study unit. Of the 204 organic constituents analyzed, 44 were detected at least once in the CLAB study unit (table 4). Of these 44 constituents, 31 have human-health benchmarks.

Twelve organic constituents were selected for additional evaluation in the status assessment because they were detected at moderate or high concentrations in the grid wells, or were detected at any concentration in greater than 10 percent of the grid wells: the solvents carbon tetrachloride, perchloroethene (PCE), trichloroethene (TCE), 1,1-dichloroethene (1,1-DCE), and cis-1,2-dichloroethene (cis-1,2-DCE); the trihalomethanes chloroform and bromodichloromethane; the gasoline additive methyl $tert$-butyl ether (MTBE); and the herbicides atrazine, simazine, prometon, and tebuthiuron (table 6, figs. 13, 14). An additional four organic constituents were selected for additional evaluation because they were reported at high or moderate concentrations in the CDPH database during the period June 4, 2003, to June 4, 2006: 1,1-dichloroethane (1,1-DCA), 1,2-dichloroethane (1,2-DCA), vinyl chloride, and total trihalomethanes (table 6).

Organic constituents as a group were present at high relative-concentrations in 3.7 percent of the primary aquifer system (spatially weighted), at moderate relative-concentrations in 13 percent, and at low relative-concentrations or not detected in 84 percent. The organic constituents selected for additional evaluation are discussed by constituent class: solvents, trihalomethanes, gasoline additives, and herbicides.

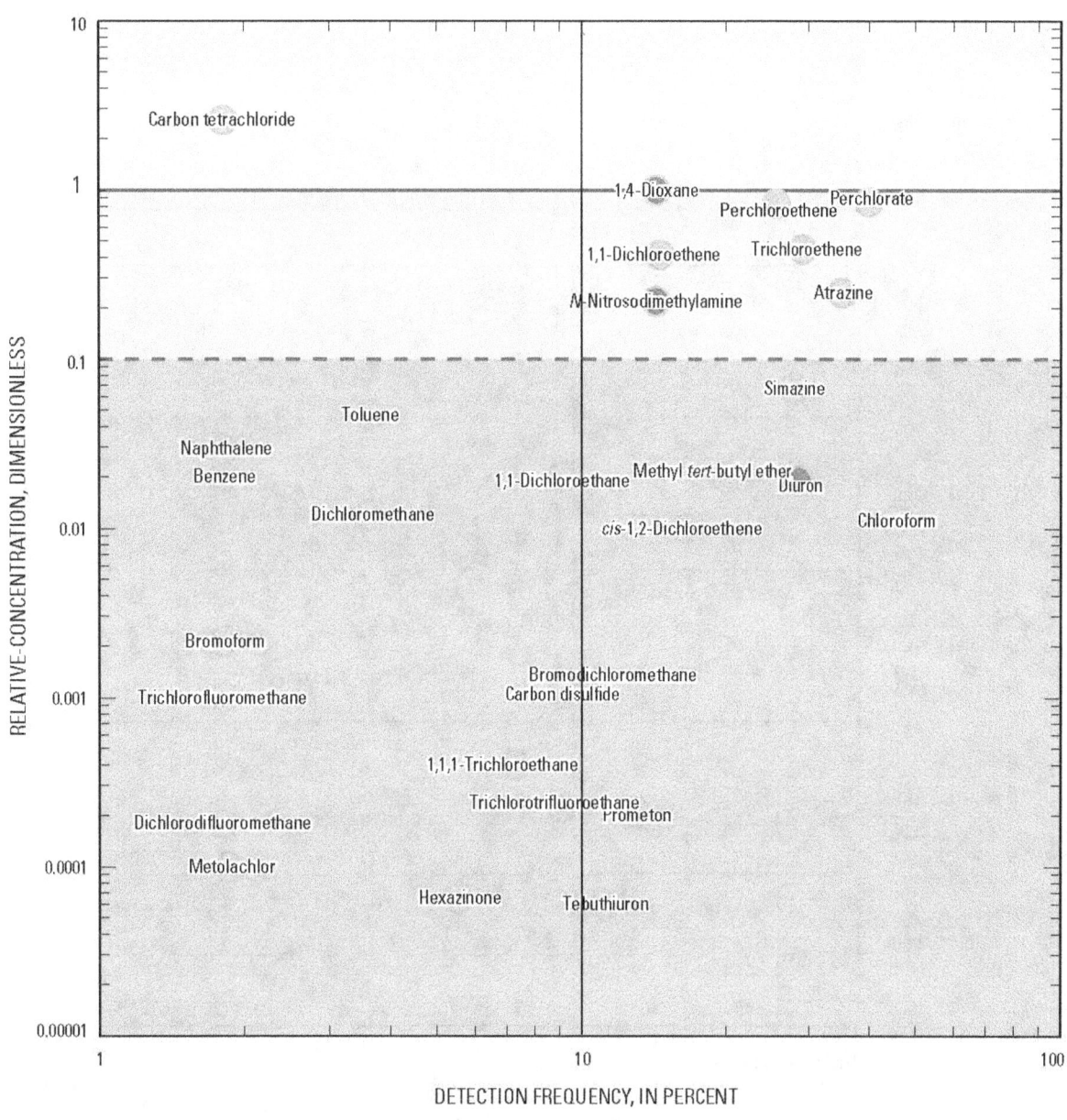

Figure 13.   Detection frequency and maximum relative-concentration for organic and special-interest constituents detected in grid wells, Coastal Los Angeles Basin study unit, 2006, California GAMA Priority Basin Project.

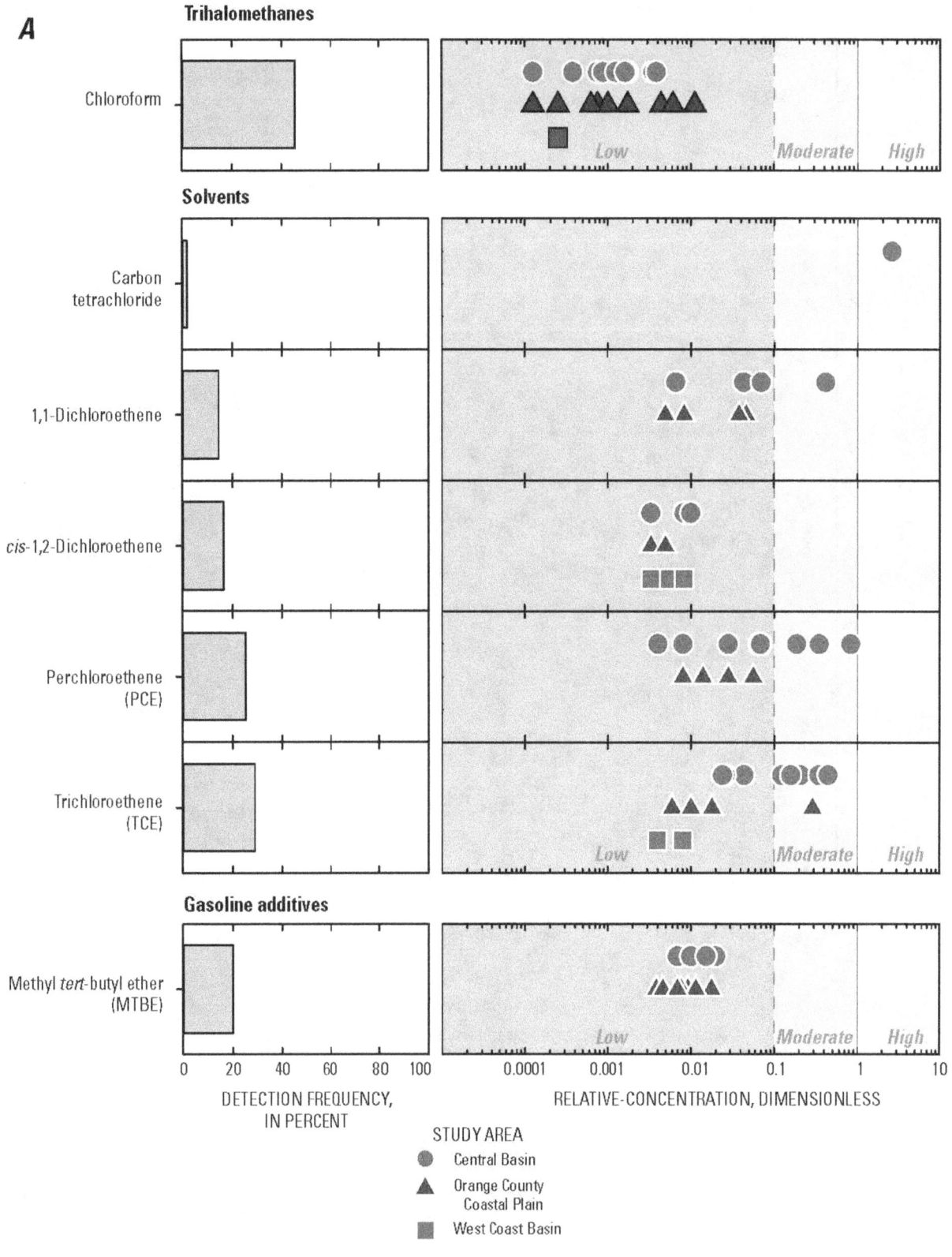

Figure 14. Detection frequencies and relative-concentrations in grid wells for selected (*A*) volatile organic compounds, and (*B*) pesticides and special-interest constituents, Coastal Los Angeles Basin study unit, 2006, GAMA Priority Basin Project.

B

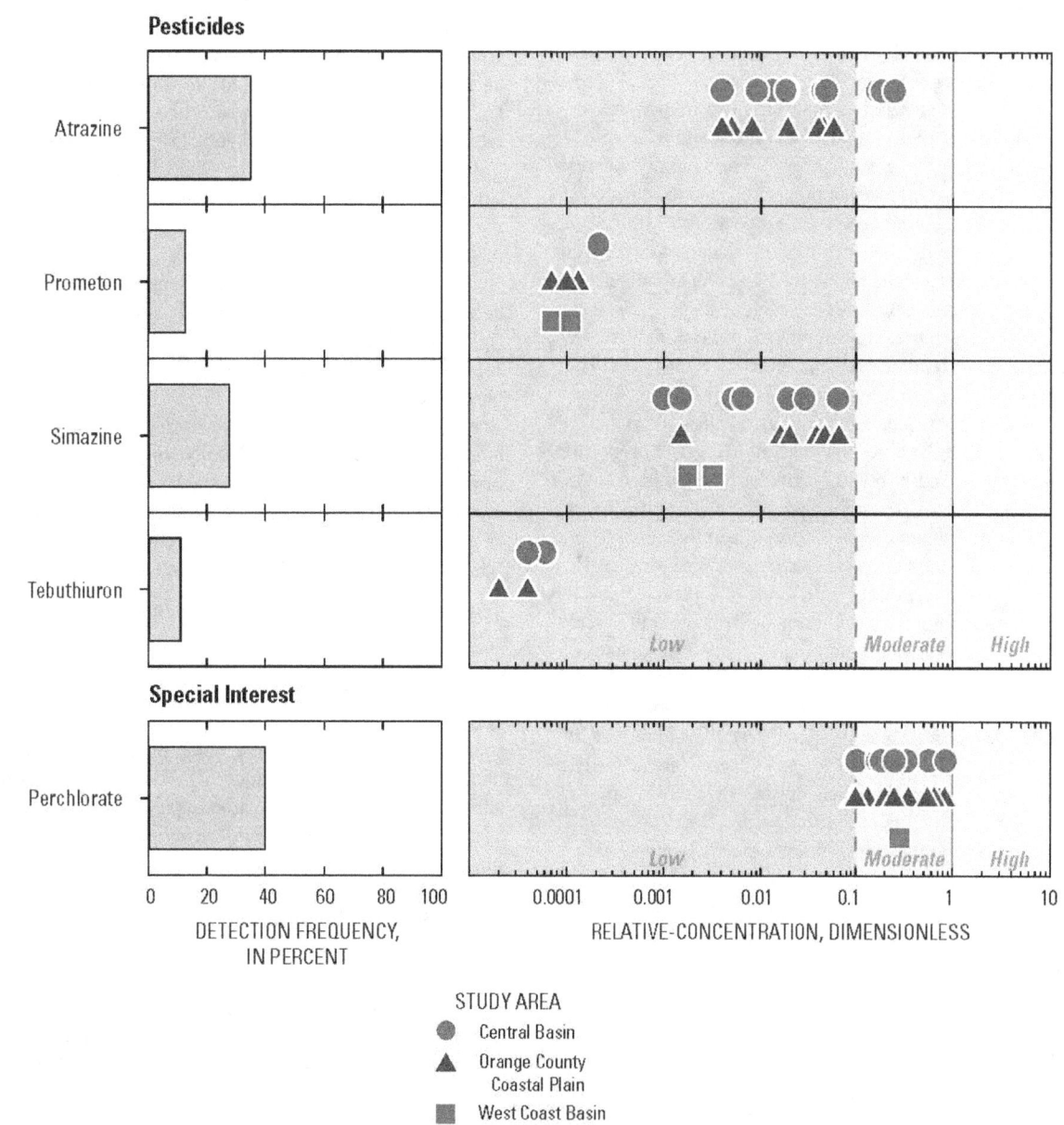

Figure 14.—Continued

Solvents

Solvents are used for various industrial, commercial, and domestic purposes. Solvents, as a class of VOCs, were present at high relative-concentrations in 3.7 percent of the primary aquifer system, and at moderate relative-concentrations in 11 percent (table 7). The spatially weighted estimate of high aquifer-scale proportion was used instead of the grid-based estimate because the detections of individual solvents at high relative-concentrations occurred in different wells. A total of 19 wells had at least 1 solvent detected at a high relative-concentration, and each of the 6 solvents present at high relative-concentrations in the study unit was detected at high relative-concentrations in 3 to 8 wells. The 19 wells with high relative-concentrations of one or more solvents were distributed across 10 grid cells. Given this distribution of high values, the high aquifer-scale proportion for solvents as a class should be greater than the high aquifer-scale proportion for any individual solvent. The grid-based calculation of high aquifer scale proportion for solvents yielded 1.8 percent, which was nearly the same as the high aquifer-scale proportion for TCE alone (1.7 percent). For this reason, the spatially weighted estimate of high aquifer-scale proportion of solvents was considered more representative than the grid-based estimate.

Most of the wells with detections of solvents at high or moderate relative-concentrations were located in the CB study area (fig. 15A). In the CB and OC study areas, wells with detections of solvents generally were located in the central and inland parts of the study areas, with fewer wells with detections on the coastal side of the study areas. This distribution of solvents reflects the dominant pattern of groundwater flow in the basins (Dawson and others, 2003). Twenty-three of the 55 grid wells (42 percent) had detections of at least 1 solvent at any concentration.

TCE and PCE were the most commonly detected solvents in the CLAB study unit, with detection frequencies of 29 percent and 25 percent, respectively (fig. 13). They also were the most commonly detected solvents in previous studies of the Coastal Los Angeles basins (Shelton and others, 2001; Dawson and others, 2003) and in a national survey of VOCs in groundwater (Zogorski and others, 2006). PCE primarily is used for dry-cleaning of fabrics and degreasing of metal parts, and is an ingredient in a wide range of products including paint removers, polishes, printing inks, lubricants, and adhesives (Doherty, 2000). TCE has similar uses as PCE, and along with *cis*-1,2-DCE, may be formed by degradation of PCE in groundwater (Vogel and McCarty, 1985). The spatially

weighted estimates of high aquifer-scale proportions were used for TCE and PCE because the grid-based proportions were zero. TCE and PCE were present at high aquifer-scale proportions (spatially weighted) in 1.7 and 1.1 percent of the primary aquifer system, respectively, and at moderate relative-concentrations in 13 and 5.5 percent, respectively (table 6). The moderate aquifer-scale proportion for solvents is less than that for TCE alone because one of the samples with a moderate relative-concentration of TCE also had a high relative-concentration of PCE.

Carbon tetrachloride, 1,1-DCE, and 1,2-DCA each were detected at high relative-concentrations in less than or equal to 1.0 percent of the primary aquifer system, and 1,1-DCA, 1,1-DCE, 1,2-DCA, *cis*-1,2-DCE, and vinyl chloride each were detected at moderate relative-concentrations in less than or equal to 1.8 percent (spatially weighted; table 6). Solvents tended to co-occur. Of the 18 wells with detections of one or more of these 6 solvents at moderate or high relative-concentrations, 13 wells (72 percent) also had detections of PCE and (or) TCE at moderate or high relative-concentrations.

Trihalomethanes

Water used for drinking water and other household uses in domestic and public (municipal and community) systems commonly is disinfected with solutions that contain chlorine. In addition to disinfecting the water, the chlorine can react with organic matter to produce THMs and other chlorinated and (or) brominated disinfection byproducts. As a class, THMs were not present at high relative-concentrations in the primary aquifer system, and were present at moderate relative-concentrations (spatially weighted) in 2.2 percent (table 7B). Chloroform was the most commonly detected VOC in the CLAB study unit, with a detection frequency of 45 percent (fig. 14B). It was the most commonly detected VOC in previous studies of the Coastal Los Angeles basins (Shelton and others, 2001; Dawson and others, 2003) and in a national survey of VOCs in groundwater (Zogorski and others, 2006).

Gasoline Additives

The gasoline oxygenate MTBE was present at moderate relative-concentrations in 0.4 percent of the primary aquifer system, based on the spatially weighted approach (table 6). MTBE was detected in more than 10 percent of the grid wells (fig. 14A).

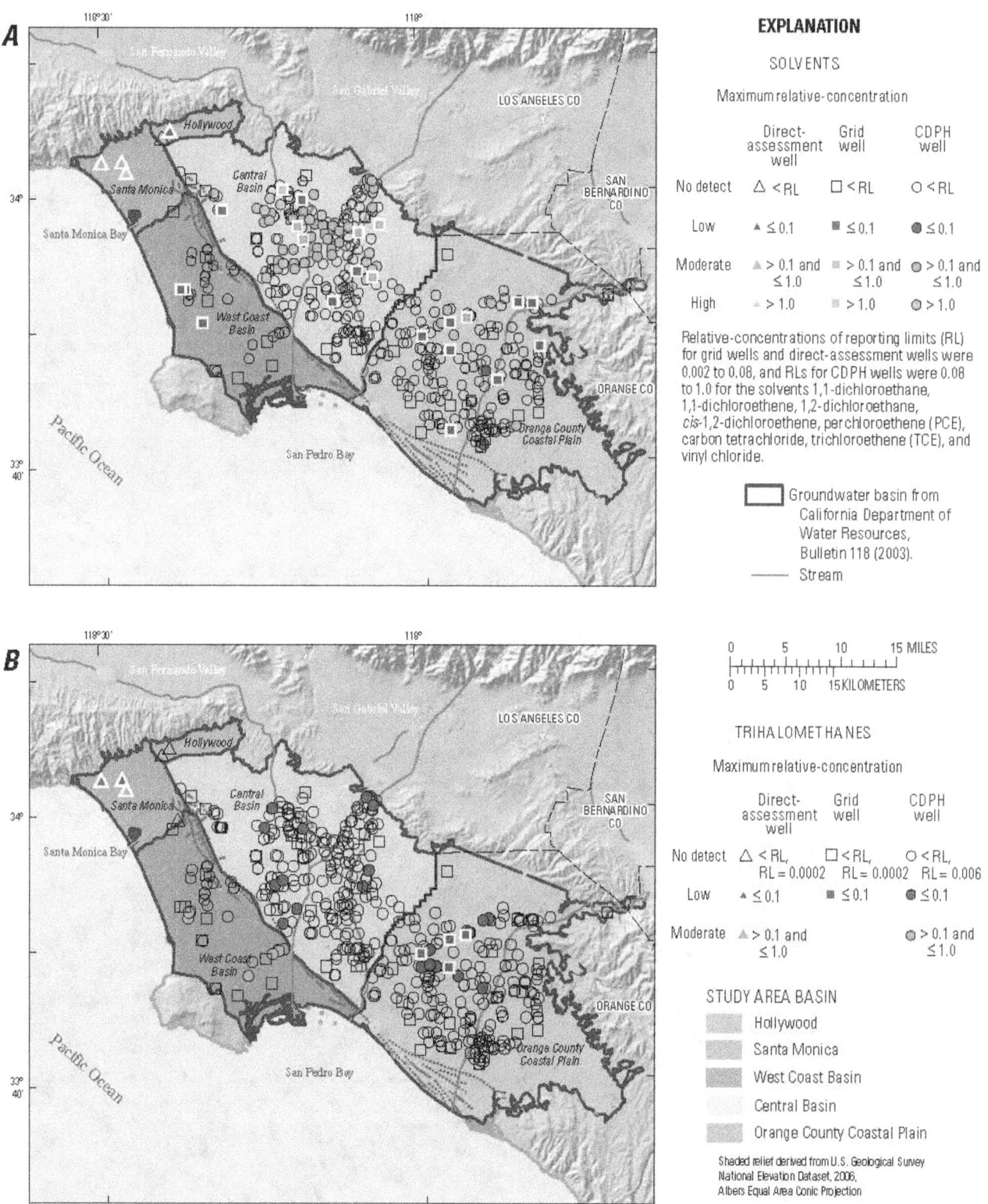

Figure 15. Maximum relative-concentration of constituents in selected organic constituent classes and special-interest constituents in grid wells and the most recent analysis, June to November 2006, for CDPH wells, Coastal Los Angeles Basin study unit, 2006, California GAMA Priority Basin Project: (*A*) solvents, (*B*) trihalomethanes, (*C*) herbicides, (*D*) perchlorate, and (*E*) 1,4-dioxane.

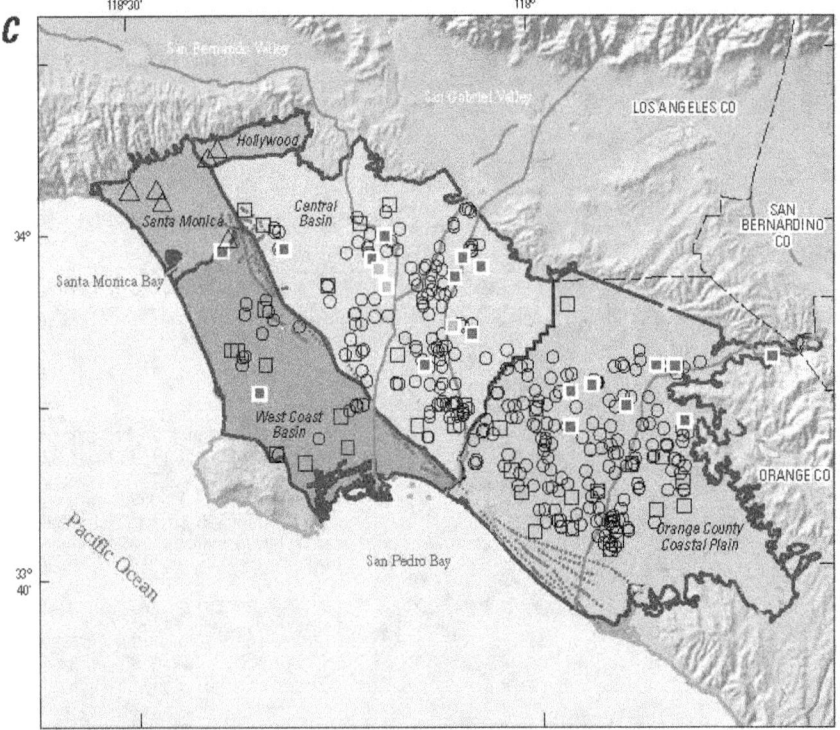

EXPLANATION

HERBICIDES

Maximum relative-concentration

	Direct-assessment well	Grid well	CDPH well
No detect	△ < RL	□ < RL	○ < RL
Low		■ ≤ 0.1	
Moderate		▨ > 0.1 and ≤ 1.0	

Relative-concentrations of reporting limits (RL) for grid wells and direct-assessment wells were 0.00002 to 0.004, and RLs for CDPH wells were 0.05 to 0.5 for the herbicides atrazine, prometon, simazine, and tebuthiuron.

0   5   10   15 MILES

0   5   10   15 KILOMETERS

Shaded relief derived from U.S. Geological Survey
National Elevation Dataset, 2006,
Albers Equal Area Conic Projection

PERCHLORATE

Relative-concentration	Measured concentration, in micrograms per liter		
	Direct-assessment well	Grid well	CDPH well
No detect	△ < RL, RL = 0.5	□ < RL, RL = 0.5	○ < RL, RL = 4.0
Low		■ ≥ 0.5 and ≤ 6.0	
Moderate	▲ > 0.6 and ≤ 6.0	▨ > 0.6 and ≤ 6.0	◉ ≥ 4.0 and ≤ 6.0
High			● > 6.0

STUDY AREA BASIN

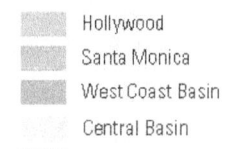

- Hollywood
- Santa Monica
- West Coast Basin
- Central Basin
- Orange County Coastal Plain

☐ Groundwater basin from California Department of Water Resources, Bulletin 118 (2003).

—— Stream

Figure 15.—Continued

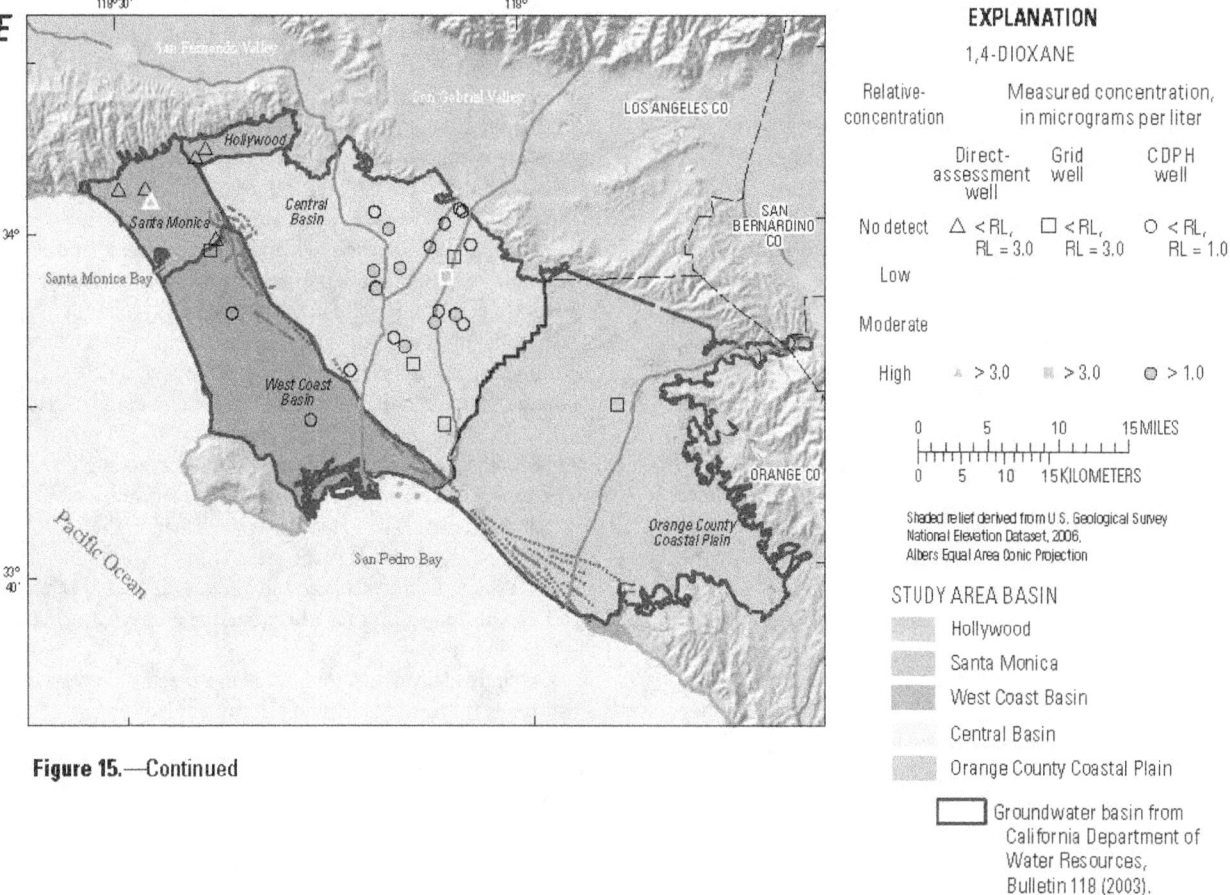

E

Figure 15.—Continued

Herbicides

As a class, herbicides were not detected at high relative-concentrations and were detected at moderate relative-concentrations in 5.5 percent of the primary aquifer system (table 7). Atrazine was the most commonly detected herbicide in grid wells, with a detection frequency of 35 percent, and was the only herbicide detected at moderate relative-concentrations (figs. 13, 14B). Simazine, prometon, and tebuthiuron had detection frequencies greater than 10 percent, and were only detected at low relative-concentrations. Diuron was analyzed for in 7 of the 55 grid wells; thus, the detection frequency for diuron shown on figure 13 may not be representative. No detections of herbicides were reported in the CDPH database; however, the reporting limits for the two most commonly analyzed herbicides, atrazine and simazine, had relative-concentrations of 0.5 and 0.25, respectively. The maximum relative-concentration of an herbicide detected by USGS-GAMA was 0.24 (fig. 14).

Atrazine, prometon, and simazine were the three most commonly detected herbicides in groundwater in urban areas and in major aquifers nationally (Gilliom and others, 2006), and also are among the most commonly detected herbicides in California groundwater (Troiano and others, 2001). In California, prometon and tebuthiuron currently generally are applied to rights-of-way and landscaping, simazine to citrus, vineyards, and nuts, and atrazine to corn and forage crops (Kegley and others, 2010), although in some parts of California, atrazine and simazine have been used on rights-of-way as well (Domagalski and Dubrovsky, 1991).

Most of the wells with detections of herbicides were located in the central and inland parts of the CB and OC study areas (fig. 15C), similar to the locations of wells with detections of solvents and (or) trihalomethanes. Of the 25 wells with detections of herbicides, 24 wells (96 percent) also had detections of solvents and (or) trihalomethanes.

Special-Interest Constituents

The special-interest group includes four chemically unrelated constituents that are of interest in California because they recently have been found in groundwater as a result of advances in analytical methods, or are considered to have the potential to reach groundwater: perchlorate, 1,2,3-trichloropropane (1,2,3-TCP), N-nitrosodimethylamine (NDMA), and 1,4-dioxane (California Department of Public Health, 2011b,c,d,e). 1,2,3-TCP was not detected in the CLAB study unit.

Perchlorate has natural and anthropogenic sources to groundwater, and concentrations greater than 1 µg/L (relative-concentration of 0.17) have a high probability of being anthropogenic in origin (Fram and Belitz, 2011b). Anthropogenic sources include solid rocket fuel, explosives, safety flares, certain fertilizers, and remobilization of naturally deposited perchlorate in unsaturated zones by irrigation recharge. Perchlorate was present at high relative-concentrations in 0.5 percent of the primary aquifer system (spatially weighted) and at moderate relative-concentrations in 35 percent (table 6). The grid-based moderate aquifer-scale proportion was higher than the spatially weighted moderate aquifer-scale proportion because of the difference in reporting limits for perchlorate between USGS-GAMA and the CDPH database. The CDPH database reporting limit was 4 µg/L, corresponding to a relative-concentration of 0.67; wells with moderate relative-concentrations between 0.50 and 0.67 would have been reported as non-detections, which may have decreased the proportion of the aquifer reported to have perchlorate present at moderate relative-concentrations. The USGS-GAMA reporting limit was 0.5 µg/L, corresponding to a relative-concentration of 0.08. Most wells with high or moderate relative-concentrations of perchlorate were located in the central parts of the CB and OC study areas (fig. 15D).

Data for 1,4-dioxane and NDMA were available for 16 and 15 grid cells, respectively, which are small numbers compared to the 57 grid cells with data for perchlorate. Therefore, the estimates of aquifer-scale proportions for 1,4-dioxane and NDMA presented on table 6 may not be representative. The available data may be used to calculate a minimum estimate of the high aquifer-scale proportion for 1,4-dioxane by assuming that the 41 cells without data have low relative-concentrations. In that case, the high aquifer-scale proportion for 1,4-dioxane would be 9 percent. 1,4-Dioxane primarily has been used as a stabilizer for solvents or as a solvent (California Department of Public Health, 2011d). Most of the wells with high relative-concentrations of 1,4-dioxane are located in the center of the CB study area (fig. 15E).

Direct-Assessment Results

The Hollywood Basin and Santa Monica Basin study areas were not divided into grid cells because the small sizes of the study areas would have resulted in too few grid cells for a robust statistical assessment and because there were relatively few CDPH wells in the basins (fig. 4). Groundwater quality in the primary aquifer system of these study areas was assessed directly rather than statistically. Four wells were sampled in the Santa Monica Basin study area, and two wells were sampled in the Hollywood Basin study area (figs. 4, A1). These direct-assessment wells were assumed to be representative of the primary aquifer system in the two study areas.

Groundwater quality for inorganic constituents was different in the Santa Monica Basin and Hollywood Basin study areas compared to the other study areas. For the inorganic constituents with human-health benchmarks, boron was the only constituent present at high relative-concentrations (one well in each study area); boron was not detected at high-relative concentrations in the other study areas. Arsenic and uranium were not present at high or moderate relative-concentrations, and nitrate was present at moderate relative-concentrations in one well (fig. 12C). For the inorganic constituents with SMCL benchmarks, all of the direct-assessment wells had moderate or high relative-concentrations of TDS, compared to about half of the wells in the other study areas (fig. 12D). The high relative-concentration of TDS in one Santa Monica Basin study area well was accompanied by high relative-concentrations of chloride and sulfate. Two Santa Monica Basin study area wells had high relative-concentrations of manganese (fig. 12E) and one of iron. One Santa Monica Basin study area well (DA-06) had high relative-concentrations of boron, manganese, iron, TDS, chloride, and sulfate.

Groundwater quality for VOCs and special-interest constituents in the Santa Monica Basin and Hollywood Basin study areas was similar to that in the other study areas. TCE, PCE, and 1,4-dioxane were detected at high relative-concentrations in one Santa Monica Basin study area well (figs. 15A,E). This same well (DA-02) had moderate relative-concentrations of chloroform, perchlorate, nitrate, 1,1-DCE, and carbon tetrachloride. Carbon tetrachloride was detected at moderate relative-concentrations in two other Santa Monica Basin study area wells, one of which also had a detection of perchlorate at moderate relative-concentration (figs. 15A,D). None of the direct-assessment wells had detections of herbicides at any concentration, compared to a 40-percent detection frequency for one or more herbicides in the other study areas.

Summary

Groundwater quality in the approximately 860-square-mile (2,227-square-kilometer) Coastal Los Angeles Basin (CLAB) study unit was investigated as part of the Priority Basin Project of the Groundwater Ambient Monitoring and Assessment (GAMA) Program. The GAMA CLAB study provides a spatially unbiased characterization of untreated groundwater quality in the primary aquifer system. The assessment is based on water-quality data collected by the U.S. Geological Survey (USGS) from 69 wells in 2006, and water-quality data from 450 wells compiled in the California Department of Public Health (CDPH) database (from the period June 4, 2003, to June 4, 2006).

The status of the current quality of the groundwater resource was assessed by using data from samples analyzed for volatile organic compounds (VOCs), pesticides, and naturally occurring inorganic constituents, such as major ions and trace elements. The status assessment characterizes the quality of groundwater resources in the primary aquifer system of the CLAB study unit, not the treated drinking water delivered to consumers by water purveyors.

Relative-concentrations (sample concentration divided by the health or aesthetic-based benchmark concentration) were used for evaluating groundwater quality for those constituents that have Federal and (or) California regulatory or non-regulatory benchmarks for drinking-water quality.

Aquifer-scale proportion was used as the primary metric for evaluating regional-scale groundwater quality. High aquifer-scale proportion is defined as the percentage of the primary aquifer system with relative-concentration greater than 1.0 for a particular constituent or class of constituents; proportion is based on an areal rather than a volumetric basis. Moderate and low aquifer-scale proportions were defined as the percentage of the primary aquifer system with moderate and low relative-concentrations, respectively. Two statistical approaches, grid-based and spatially weighted, were used to evaluate aquifer-scale proportions for individual constituents and classes of constituents. Grid-based and spatially weighted estimates were comparable in the CLAB study unit (within 90-percent confidence intervals for most constituents). However, the spatially weighted approach was superior to the grid-based proportion when a constituent was present at a high relative-concentration in a small fraction of the aquifer.

Inorganic constituents with human-health benchmarks were present at high relative-concentrations in 5.5 percent of the primary aquifer system and at moderate relative concentrations in 26 percent. The high aquifer-scale proportion of inorganic constituents primarily reflected high aquifer-scale proportions of arsenic (1.9 percent), nitrate (1.9 percent), and uranium (1.2 percent). The inorganic constituents with secondary maximum contaminant levels were present at high relative-concentrations in 18 percent of the primary aquifer system and at moderate relative-concentrations in 47 percent. The high aquifer-scale proportion primarily reflected high aquifer-scale proportions of total dissolved solids (1.9 percent), manganese (15 percent), and iron (9.4 percent).

Relative-concentrations of organic constituents (one or more) were high in 3.7 percent, and moderate in 13 percent, of the primary aquifer system. The high aquifer-scale proportion of organic constituents primarily reflected high aquifer-scale proportions of solvents, including trichloroethene (TCE; 1.7 percent), perchloroethene (PCE; 1.1 percent) and carbon tetrachloride (1.0 percent). Of the 204 organic constituents analyzed, 44 constituents were detected. Eleven organic constituents had detection frequencies of greater than 10 percent: the trihalomethanes chloroform and bromodichloromethane; the solvents TCE, PCE, cis-1,2- dichloroethene, and 1,1-dichloroethene; the herbicides atrazine, simazine, prometon, and tebuthiuron;

and the gasoline additive methyl *tert*-butyl ether. Most detections were at low relative-concentrations. The special-interest constituent perchlorate was detected at high relative-concentrations in 0.5 percent of the primary aquifer system and at moderate relative-concentrations in 35 percent. The special-interest constituent 1,4-dioxane was detected at high relative-concentrations, but an insufficient number of samples was analyzed to provide a representative estimate of aquifer-scale proportion.

Acknowledgments

The authors thank the following cooperators for their support: the California State Water Board, Lawrence Livermore National Laboratory, California Department of Public Health, and California Department of Water Resources. We especially thank the well owners and water purveyors for their cooperation in allowing the U.S. Geological Survey to collect samples from their wells. Funding for this work was provided by State of California bonds authorized by Proposition 50 and administered by the California State Water Board.

References Cited

Belitz, K., Dubrovsky, N.M., Burow, K.R., Jurgens, B., and Johnson, T., 2003, Framework for a ground-water quality monitoring and assessment program for California: U.S. Geological Survey Water-Resources Investigations Report 03-4166, 78 p.

Belitz, Kenneth, Hamlin, S.N., Burton, C.A., Kent, Robert, Fay, R.G., and Johnson, Tyler, 2004, Water quality in the Santa Ana Basin, California, 1999–2001: U.S. Geological Survey Circular 1238, 37 p.

Belitz, K., Jurgens, B., Landon, M.K., Fram, M.S., and Johnson, T., 2010, Estimation of aquifer-scale proportion using equal-area grids—Assessment of regional-scale groundwater quality: Water Resources Research, v. 46, W11550, 14 p., doi:10.1029/2010WR009321. (Also available at http://www.agu.org/pubs/crossref/2010/2010WR009321.shtml.)

Brown, L.D., Cai, T.T., and DasGupta, A., 2001, Interval estimation for a binomial proportion: Statistical Science, v. 16, no. 2, p. 101–117.

California Department of Public Health, 2010, Drinking water notification levels—Notification levels: California Department of Public Health, accessed January 10, 2012, at http://www.cdph.ca.gov/certlic/drinkingwater/Pages/NotificationLevels.aspx.

California Department of Public Health, 2011a, California drinking water-related laws—Drinking water-related regulations, Title 22: California Department of Public Health, accessed January 10, 2012, at http://www.cdph.ca.gov/certlic/drinkingwater/Pages/Lawbook.aspx.

California Department of Public Health, 2011b, Perchlorate in drinking water: California Department of Public Health, accessed January 10, 2012, at http://www.cdph.ca.gov/certlic/drinkingwater/Pages/Perchlorate.aspx.

California Department of Public Health, 2011c, NDMA and other nitrosamines—Drinking water issues: California Department of Public Health, accessed January 10, 2012, at http://www.cdph.ca.gov/certlic/drinkingwater/Pages/NDMA.aspx.

California Department of Public Health, 2011d, 1,4-Dioxane: California Department of Public Health, accessed January 10, 2012, at http://www.cdph.ca.gov/certlic/drinkingwater/Pages/1,4-Dioxane.aspx.

California Department of Public Health, 2011e, 1,2,3-Trichloropropane: California Department of Public Health, accessed January 10, 2012, at http://www.cdph.ca.gov/certlic/drinkingwater/Pages/123TCP.aspx.

California Department of Water Resources, 2003, California's groundwater: California Department of Water Resources Bulletin 118, 246 p., accessed June 20, 2007, at http://www.water.ca.gov/groundwater/bulletin118/bulletin118update2003.cfm.

California State Water Resources Control Board, 2001, Geographic Environmental Information Management System GeoTracker (GEIMS) Leaking Underground Fuel/Storage Tank database (LUFT) [digital data]: Sacramento, California, California Environmental Protection Agency, State Water Resources Control Board, Division of Water Quality.

Chapelle, F.H., 2001, Ground-water microbiology and geochemistry (2nd ed.): New York, John Wiley and Sons, Inc., 477 p.

Chapelle, F.H., McMahon, P.B., Dubrovsky, N.M., Fuji, R.F., Oaksford, E.T., and Vroblesky, D.A., 1995, Deducing the distribution of terminal electron-accepting processes in hydrologically diverse groundwater systems: Water Resources Research, v. 31, no. 2, p. 359–371.

Clark, I.D., and Fritz, P., 1997, Environmental isotopes in hydrogeology: New York, Lewis Publishers, 328 p.

Cook, P.G., and Böhlke, J.K., 2000, Determining timescales for groundwater flow and solute transport, in Cook, P.G., and Herczeg, A., eds., Environmental tracers in subsurface hydrology: Boston, Kluwer Academic Publishers, p. 1–30.

Craig, H., and Lal, D., 1961, The production rate of natural tritium: Tellus, v. 13, p. 85–105.

Dawson, B.J.M., Belitz, Kenneth, Land, Michael, and Danskin, W.R., 2003, Stable isotopes and volatile organic compounds along seven ground-water flow paths in divergent and convergent flow systems, Southern California, 2000: U.S. Geological Survey Water-Resources Investigations Report 03-4059, 79 p.

Doherty, R.E., 2000, A history of the production and use of carbon tetrachloride, tetrachloroethylene, trichloroethylene, and 1,1,1-trichloroethane in the United States. Part 1—historical background; carbon tetrachloride and tetrachloroethylene: Journal of Environmental Forensics, v. 1, p. 69–81.

Domagalski, J.L., and Dubrovsky, N.M., 1991, Regional assessment of nonpoint-source pesticide residues in ground water, San Joaquin Valley, California: U.S. Geological Survey Water-Resources Investigations Report 91-4027, 14 p.

Focazio, M.J., Welch, A.H., Watkins, S.A., Helsel, D.R., and Horn, M.A., 1999, A retrospective analysis on the occurrence of arsenic in groundwater resources of the United States and limitations in drinking water supply characterizations: U.S. Geological Survey Water-Resources Investigations Report 99-4279, 21 p.

Fontes, J.C., and Garnier, J.M., 1979, Determination of the initial ^{14}C activity of the total dissolved carbon—A review of the existing models and a new approach: Water Resources Research, v. 15, no. 2, p. 399–413.

Fram, M.S., and Belitz, K., 2011a, Occurrence and concentrations of pharmaceutical compounds in groundwater used for public drinking-water supply in California: Science of the Total Environment, v. 409, p. 3409–3417.

Fram, M.S., and Belitz, K., 2011b, Probability of detecting perchlorate under natural conditions in deep groundwater in California and the southwestern United States: Environmental Science & Technology, v. 45, no. 4, p. 1271–1277.

Fram, M.S., and Belitz, Kenneth, 2012, Status and understanding of groundwater quality in the Tahoe-Martis, Central Sierra, and Southern Sierra study units, 2006–2007—California GAMA Priority Basin Project: U.S. Geological Survey Scientific Investigations Report 2011-5216, 222 p.

Gilliom, R.J., Barbash, J.E., Crawford, C.G., Hamilton, P.A., Martin, J.D., Nakagaki, N., Nowell, L.H., Scott, J.C., Stackelberg, P.E., Thelin, G.P., and Wolock, D.M., 2006, The quality of our nation's waters—Pesticides in the nation's streams and ground water, 1992–2001: U.S. Geological Survey Circular 1291, 172 p.

Hamlin, S.N., Belitz, Kenneth, and Johnson, Tyler, 2005, Occurrence and distribution of volatile organic compounds and pesticides in ground water in relation to hydrologic characteristics and land use in the Santa Ana Basin, Southern California: U.S. Geological Survey Scientific Investigations Report 2005-5032, 40 p.

Hamlin, S.N., Belitz, Kenneth, Kraja, Sarah, and Dawson, Barbara, 2002, Groundwater quality in the Santa Ana Watershed, California—Overview and data summary: U.S. Geological Survey Water-Resources Investigations Report 02-4243, 137 p.

Hem, J.D., 1985, Study and interpretation of the chemical characteristics of natural water (3d ed.): U.S. Geological Survey Water-Supply Paper 2254, 213 p.

Herndon, R.L., Brukner, D.B., and Sharp, G., 1997, Groundwater systems in the Orange County groundwater basin, Phase 1A Task 2.2 Report, prepared for the Santa Ana Watershed Project Authority, TIN/TOS Task Force: Orange County Water District, 12 p.

Isaaks, E.H., and Srivastava, R.M., 1989, Applied geostatistics: New York, Oxford University Press, 561 p.

Johnson, T.D., and Belitz, K., 2009, Assigning land use to supply wells for the statistical characterization of regional groundwater quality—Correlating urban land use and VOC occurrence: Journal of Hydrology, v. 370, p. 100–108.

Jurgens, B.C., Fram, M.S., Belitz, K., Burow, K.R., and Landon, M.K., 2010, Effects of groundwater development on uranium—Central Valley, USA: Ground Water, v. 48, no. 6, p. 913–928.

Jurgens, B.C., McMahon, P.B., Chapelle, F.H., and Eberts, S.M., 2009, An Excel® workbook for identifying redox processes in ground water: U.S. Geological Survey Open-File Report 2009-1004, 8 p. (Also available at http://pubs.usgs.gov/of/2009/1004/.)

Kulongoski, J.T., and Belitz, Kenneth, 2004, Groundwater ambient monitoring and assessment program: U.S. Geological Survey Fact Sheet 2004-3088, 4 p. (Also available at http://pubs.usgs.gov/fs/2004/3088/.)

Kulongoski, J.T., Belitz, Kenneth, Landon, M.K., and Farrar, Christopher, 2010, Status and understanding of groundwater quality in the North San Francisco Bay groundwater basins, 2004: California GAMA Priority Basin Project: U.S. Geological Survey Scientific Investigations Report 2010-5089, 87 p. (Also available at http://pubs.usgs.gov/sir/2010/5089.)

Landon, M.K., Belitz, Kenneth, Jurgens, B.C., Kulongoski, J.T., and Johnson, T.D., 2010, Status and understanding of groundwater quality in the Central-Eastside San Joaquin Basin, 2006—California GAMA Priority Basin Project: U.S. Geological Survey Scientific Investigations Report 2009-5266, 97 p. (Also available at http://pubs.usgs.gov/sir/2009/5266/.)

Lucas, L.L., and Unterweger, M.P., 2000, Comprehensive review and critical evaluation of the half-life of tritium: Journal of Research of the National Institute of Standards and Technology, v. 105, no. 4, p. 541–549.

Mathany, T.M., Land, Michael, and Belitz, Kenneth, 2008, Ground-water quality data in the Coastal Los Angeles Basin study unit, 2006—Results from the California GAMA Program: U.S. Geological Survey Data Series 387, 98 p. (Also available at http://pubs.usgs.gov/ds/387/.)

McMahon, P.B., and Chapelle, F.H., 2008, Redox processes and water quality of selected principal aquifer systems: Ground Water, v. 46, no. 2, p. 259–271.

Michel, R.L., 1989, Tritium deposition in the continental United States, 1953–83: U.S. Geological Survey Water-Resources Investigations Report 89-4072, 46 p.

Michel, R.L., and Schroeder, R., 1994, Use of long-term tritium records from the Colorado River to determine timescales for hydrologic processes associated with irrigation in the Imperial Valley, California: Applied Geochemistry, v. 9, p. 387–401.

Nakagaki, Naomi, and Wolock, D.M., 2005, Estimation of agricultural pesticide use in drainage basins using land cover maps and county pesticide data: U.S. Geological Survey Open-File Report 2005-1188, 46 p.

Nakagaki, Naomi, Price, C.V., Falcone, J.A., Hitt, K.J., and Ruddy, B.C., 2007, Enhanced National Land Cover Data 1992 (NLCDe 92): U.S. Geological Survey Raster digital data, available online at http://water.usgs.gov/lookup/getspatial?nlcde92.

National Oceanic and Atmospheric Administration, 2007, The climate of Los Angeles, California, accessed March 2007 at http://www.wrh.noaa.gov/lox/archive/pns_2007summary.pdf.

Piper, A.M., 1944, A graphic procedure in the geochemical interpretation of water analyses: American Geophysical Union Transactions, v. 25, p. 914–923.

Plummer, L.N., Michel, R.L., Thurman, E.M., and Glynn, P.D., 1993, Environmental tracers for age-dating young ground water, in Alley, W.M., ed., Regional ground-water quality: New York, Van Nostrand Reinhold, p. 255–294.

Rowe, B.L., Toccalino, P.L., Moran, M.J., Zogorski, J.S., and Price, C.V., 2007, Occurrence and potential human-health relevance of volatile organic compounds in drinking water from domestic wells in the United States: Environmental Health Perspectives, v. 115, no. 11, p. 1539–1546.

Scott, J.C., 1990, Computerized stratified random site selection approaches for design of a groundwater quality sampling network: U.S. Geological Survey Water-Resources Investigations Report 90-4101, 109 p.

Shelton, J.L., Burow, K.R., Belitz, Kenneth, Dubrovsky, N.M., Land, Michael, and Gronberg, J.A., 2001, Low-level volatile organic compounds in active public supply wells as ground-water tracers in the Los Angeles physiographic basin, California, 2000: U.S. Geological Survey Water-Resources Investigations Report 01-4188, 29 p.

State of California, 1999, Supplemental Report of the 1999 Budget Act 1999–00 Fiscal Year, Item 3940-001-0001, State Water Resources Control Board, accessed September 9, 2010, at http://www.lao.ca.gov/1999/99-00_supp_rpt_lang. html#3940.

State of California, 2001a, Assembly Bill No. 599, Chapter 522, accessed September 9, 2010, at http://www. swrcb.ca.gov/gama/docs/ab_599_bill_20011005_chaptered. pdf.

State of California, 2001b, Groundwater Monitoring Act of 2001: California Water Code, part 2.76, Sections 10780–10782.3, accessed September 9, 2010, at http:// www.leginfo.ca.gov/cgi-bin/displaycode?section=wat&gr oup=10001-11000&file=10780-10782.3.

State Water Resources Control Board, 2003, A comprehensive groundwater quality monitoring program for California: Assembly Bill 99 Report to the Governor and Legislature, March 2003, 100 p. http://www.waterboards.ca.gov/ water_issues/programs/gama/docs/final_ab_599_rpt_to_ legis_7_31_03.pdf.

Toccalino, P.L., and Norman, J.E., 2006, Health-based screening levels to evaluate U.S. Geological Survey groundwater quality data: Risk Analysis, v. 26, no. 5, p. 1339–1348.

Toccalino, P.L., Norman, J., Phillips, R., Kauffman, L., Stackelberg, P., Nowell, L., Krietzman, S., and Post, G., 2004, Application of health-based screening levels to groundwater quality data in a state-scale pilot effort: U.S. Geological Survey Scientific Investigations Report 2004-5174, 14 p.

Tolstikhin, I.N., and Kamensky, I.L., 1969, Determination of groundwater ages by the T-^{3}He method: Geochemistry International, v. 6, p. 810–811.

Troiano, J., Weaver, D., Marade, J., Spurlock, F., Pepple, M., Nordmark, C., and Bartkowiak, D., 2001, Summary of well water sampling in California to detect pesticide residues resulting from nonpoint source applications: Journal of Environmental Quality, v. 30, p. 448–459.

U.S. Census Bureau, 1990, Census of population and housing, summary tape file 3A, accessed August 11, 2011, at ftp:// ftp2.census.gov/census_1990/.

U.S. Environmental Protection Agency, 1998, Code of Federal Regulations, title 40—protection of environment, chapter 1—environmental protection agency, subchapter E— pesticide programs, part 159—statements of policies and interpretations, subpart D—reporting requirements for risk/benefit information, 40 CFR 159.184: Washington, National Archives and Records Administration, September 19, 1997; amended June 19, 1998, accessed September 5, 2008, at http://www.epa.gov/EPA-PEST/1997/ September/Day-19/p24937.htm.

U.S. Environmental Protection Agency, 1999, Proposed radon in drinking water rule, accessed January 10, 2012, at http:// water.epa.gov/lawsregs/rulesregs/sdwa/radon/regulations. cfm.

U.S. Environmental Protection Agency, 2009, Drinking water contaminants, accessed January 10, 2012, at http://www. epa.gov/safewater/contaminants/index.html.

U.S. Environmental Protection Agency, 2011, Drinking water health advisories—2011 Drinking water standards and health advisory tables, accessed January 10, 2012, at http:// www.epa.gov/waterscience/criteria/drinking/.

United Nations Educational, Scientific, and Cultural Organization (UNESCO), 1979, Map of the world distribution of arid regions—Explanatory note. MAB Technical Notes, v. 7, 42 p.

Vogel, J.C., and Ehhalt, D., 1963, The use of the carbon isotopes in groundwater studies, in Radioisotopes in Hydrology: Tokyo, IAEA, p. 383–395.

Vogel, T.M., and McCarty, P.L., 1985, Biotransformation of tetrachloroethylene to trichloroethylene, dichloroethylene, vinyl chloride, and carbon dioxide under methanogenic conditions: Applied and Environmental Microbiology, v. 49, no. 5, p. 1080–1083.

Welch, A.H., Westjohn, D.B., Helsel, D.R., and Wanty, R.B., 2000, Arsenic in groundwater of the United States— Occurrence and geochemistry: Ground Water, v. 38, no. 4, p. 589–604.

Zogorski, J.S., Carter, J.M., Ivahnenko, T., Lapham, W.W., Moran, M.J., Rowe, B.L., Squillace, P.J., and Toccalino, P.L., 2006, Volatile organic compounds in the Nation's ground water and drinking-water supply wells: U.S. Geological Survey Circular 1292, 101 p.

Appendix A. Use of Data From the California Department of Public Health (CDPH) Database

In the CLAB study unit, the historical CDPH database contains more than 502,000 records distributed across more than 850 wells, requiring targeted retrievals to manageably use the data to assess water quality. The following paragraphs summarize the selection process for wells and data from the CDPH database for use in the grid-based status assessment.

The grid-based calculation of aquifer-scale proportion uses one value per grid cell. Where USGS data for inorganic constituents were not available, additional data to represent a cell were selected in the following order of priority: (1) analyses made by USGS-GAMA sampling from the USGS-grid well, (2) data obtained from the CDPH database for the USGS-grid well, (3) data obtained from the CDPH database for another well in the cell. Of the 61 grid cells in the CLAB study unit, 3 cells had USGS-grid wells with the full complement of inorganic constituent data collected by USGS-GAMA, 5 cells had USGS-grid wells with USGS data for all inorganic constituents except for radioactive constituents, 47 cells had USGS-grid wells with no USGS data for inorganic constituents, and 6 cells had no USGS-grid wells. The CDPH database was queried to provide these missing data for inorganic constituents. CDPH wells with data for the most recent 3 years available at the time of sampling (June 4, 2003, through June 4, 2006) were considered. If a well had more than one analysis for a constituent in the 3-year interval, then the most recent data were selected.

The data in the CDPH database is of unknown quality, and the database does not contain data for quality-control samples with which to make a comprehensive quality-control assessment of the data. Cation-anion imbalance was used as a rough quality-control metric. Because water is electrically neutral and must have a balance between positive (cations) and negative (anions) electrically charged dissolved species, the cation/anion imbalance commonly is used as a quality-assurance check for water-sample analysis (Hem, 1985). An imbalance of less than 10 percent was defined as indicating acceptable quality of the major ion data for the sample. It was assumed that if the quality of the major ion data were acceptable, then the quality of the data for the other inorganic constituents also would be acceptable. In practice, however, some wells did not have data for major ion constituents, so the cation-anion imbalance could not be evaluated.

Fifty-two cells were represented by a USGS-grid well lacking data for all or some inorganic constituents, and in 38 of those cells, the CDPH database contained data for the USGS-grid well for all or some of the missing inorganic constituents. These wells were assigned well identification numbers consisting of the same prefix and number of the USGS-grid well with the extra prefix DG inserted before

the number (table A1; fig. A1B). This procedure generally yielded data for some, but not all, of the missing inorganic constituents.

If the CDPH-grid well that was the same well as the USGS-grid well did not provide all of the missing data to represent the cell, or the USGS-grid well was not a well with data in the CDPH database, or the cell did not contain a USGS-grid well, a well was selected from the CDPH database to provide the missing data for the cell. This well was selected on the basis of randomized rank, cation-anion imbalance, and presence of data. The highest ranking well with acceptable cation-anion balance and with data for the most of the missing constituents was selected. If no wells had acceptable cation-anion balances, or if the wells with acceptable cation-anion balances lacked data for the missing constituents, then the highest ranking well with data for the most of the missing constituents was selected. Thirty-nine wells were selected by this procedure. These wells were assigned well identification numbers consisting of the same prefix and number of the USGS-grid well with the extra prefix DPH inserted before the number (table A1; fig. A1C). No more than one CDPH-grid well was selected in a cell using this procedure. CDPH-grid wells in cells with no USGS-grid wells were labeled with the next number in the sequence. The combination of the USGS-grid wells and CDPH-grid wells produced a grid-well network covering 57 of the 61 grid cells in the CLAB study unit (table A1). No accessible wells or data were available for the remaining four cells.

The result of these steps was a dataset for inorganic constituents, with one value for each cell for each constituent, having data from the USGS database, the CDPH database, or both databases, for 57 cells. Because the CDPH database did not contain data for all of the missing inorganic constituents in all of the cells, the number of cells represented by values varied between the constituents from a minimum of 10 for strontium to a maximum of 56 for specific conductance (table 2). Most of the inorganic constituents had values representing at least 50 of the cells. Estimates of aquifer-scale proportion for constituents based on a smaller number of wells are subject to a larger error associated with the 90-percent confidence intervals (on the basis of Jeffreys interval for the binomial distribution).

Differences in constituent reporting levels associated with USGS and CDPH data did not affect analysis of high or moderate relative-concentrations because concentrations greater than one-half of water-quality benchmarks were substantially higher than the reporting levels. Several types of comparisons between USGS-collected data and CDPH data are described in appendix D.

Table A1.   Nomenclature for USGS-grid and CDPH-grid wells used in the status assessment, Coastal Los Angeles Basin study unit, 2006, California GAMA Priority Basin Project.

[CDPH-grid wells are labeled "DG" if the USGS-grid and CDPH-grid wells are the same well, and are labeled "DPH" if the CDPH-grid and USGS-grid wells are different wells, or if no USGS-grid well exists in the cell. The most recent data from the CDPH database for the time period June 4, 2003, to June 4, 2006, for the "DG" and, secondarily, "DPH" CDPH-grid wells were used to provide data for inorganic constituents for grid cells lacking USGS-GAMA data for those constituents. Two of the "DPH" CDPH-grid wells were the same wells as two USGS-understanding wells; data collected by USGS-GAMA were preferred. Abbreviations: CDPH, California Department of Public Health; USGS, U.S. Geological Survey; –, no well sampled or selected]

Well identification number		
USGS-grid	"DG" CDPH-grid	"DPH" CDPH-grid
Central Basin study area		
CLABCB-01	CLABCB-DG-01	–
CLABCB-02	CLABCB-DG-02	CLABCB-DPH-02
CLABCB-03	–	–
CLABCB-04	–	–
CLABCB-05	–	CLABCB-DPH-05
CLABCB-06	CLABCB-DG-06	CLABCB-DPH-06
CLABCB-07	CLABCB-DG-07	CLABCB-DPH-07
CLABCB-08	CLABCB-DG-08	CLABCB-DPH-08
CLABCB-09	CLABCB-DG-09	CLABCB-DPH-09
CLABCB-10	CLABCB-DG-10	CLABCB-DPH-10
CLABCB-11	CLABCB-DG-11	CLABCB-DPH-11
CLABCB-12	CLABCB-DG-12	CLABCB-DPH-12
CLABCB-13	CLABCB-DG-13	CLABCB-DPH-13[1]
CLABCB-14	CLABCB-DG-14	CLABCB-DPH-14
CLABCB-15	CLABCB-DG-15	CLABCB-DPH-15
CLABCB-16	CLABCB-DG-16	CLABCB-DPH-16
CLABCB-17	CLABCB-DG-17	–
CLABCB-18	–	CLABCB-DPH-18
CLABCB-19	–	CLABCB-DPH-19
CLABCB-20	–	CLABCB-DPH-20
CLABCB-21	–	CLABCB-DPH-21
Orange County Coastal Plain study area		
CLABOC-01	CLABOC-DG-01	CLABOC-DPH-01
CLABOC-02	–	–
CLABOC-03	CLABOC-DG-03	CLABOC-DPH-03
CLABOC-04	CLABOC-DG-04	CLABOC-DPH-04
CLABOC-05	CLABOC-DG-05	–
CLABOC-06	–	CLABOC-DPH-06
CLABOC-07	CLABOC-DG-07	CLABOC-DPH-07[2]

Table A1.   Nomenclature for USGS-grid and CDPH-grid wells used in the status assessment, Coastal Los Angeles Basin study unit, 2006, California GAMA Priority Basin Project.—Continued

[CDPH-grid wells are labeled "DG" if the USGS-grid and CDPH-grid wells are the same well, and are labeled "DPH" if the CDPH-grid and USGS-grid wells are different wells, or if no USGS-grid well exists in the cell. The most recent data from the CDPH database for the time period June 4, 2003, to June 4, 2006, for the "DG" and, secondarily, "DPH" CDPH-grid wells were used to provide data for inorganic constituents for grid cells lacking USGS-GAMA data for those constituents. Two of the "DPH" CDPH-grid wells were the same wells as two USGS-understanding wells; data collected by USGS-GAMA were preferred. Abbreviations: CDPH, California Department of Public Health; USGS, U.S. Geological Survey; –, no well sampled or selected]

Well identification number		
USGS-grid	"DG" CDPH-grid	"DPH" CDPH-grid
Orange County Coastal Plain study area—Continued		
CLABOC-08	CLABOC-DG-08	CLABOC-DPH-08
CLABOC-09	CLABOC-DG-09	–
CLABOC-10	CLABOC-DG-10	–
CLABOC-11	CLABOC-DG-11	CLABOC-DPH-11
CLABOC-12	CLABOC-DG-12	CLABOC-DPH-12
CLABOC-13	CLABOC-DG-13	CLABOC-DPH-13
CLABOC-14	CLABOC-DG-14	CLABOC-DPH-14
CLABOC-15	CLABOC-DG-15	CLABOC-DPH-15
CLABOC-16	CLABOC-DG-16	–
CLABOC-17	CLABOC-DG-17	CLABOC-DPH-17
CLABOC-18	–	CLABOC-DPH-18
CLABOC-19	CLABOC-DG-19	–
CLABOC-20	CLABOC-DG-20	CLABOC-DPH-20
CLABOC-21	CLABOC-DG-21	–
CLABOC-22	CLABOC-DG-22	CLABOC-DPH-22
CLABOC-23	CLABOC-DG-23	CLABOC-DPH-23
CLABOC-24	–	CLABOC-DPH-24
West Coast Basin study area		
CLABWB-01	CLABWB-DG-02	–
CLABWB-02	CLABWB-DG-02	–
CLABWB-03	–	–
CLABWB-04	–	–
CLABWB-05	CLABWB-DG-02	CLABWB-DPH-05
CLABWB-06	–	–
CLABWB-07	–	CLABWB-DPH-07
CLABWB-08	–	CLABWB-DPH-08
CLABWB-09	CLABWB-DG-02	–
CLABWB-10	–	–
–	–	CLABWB-DPH-11
–	–	CLABWB-DPH-12

[1]CLABCB-DPH-13 is the same well as CLABU-02.

[2]CLABOC-DPH-07 is the same well as CLABU-07.

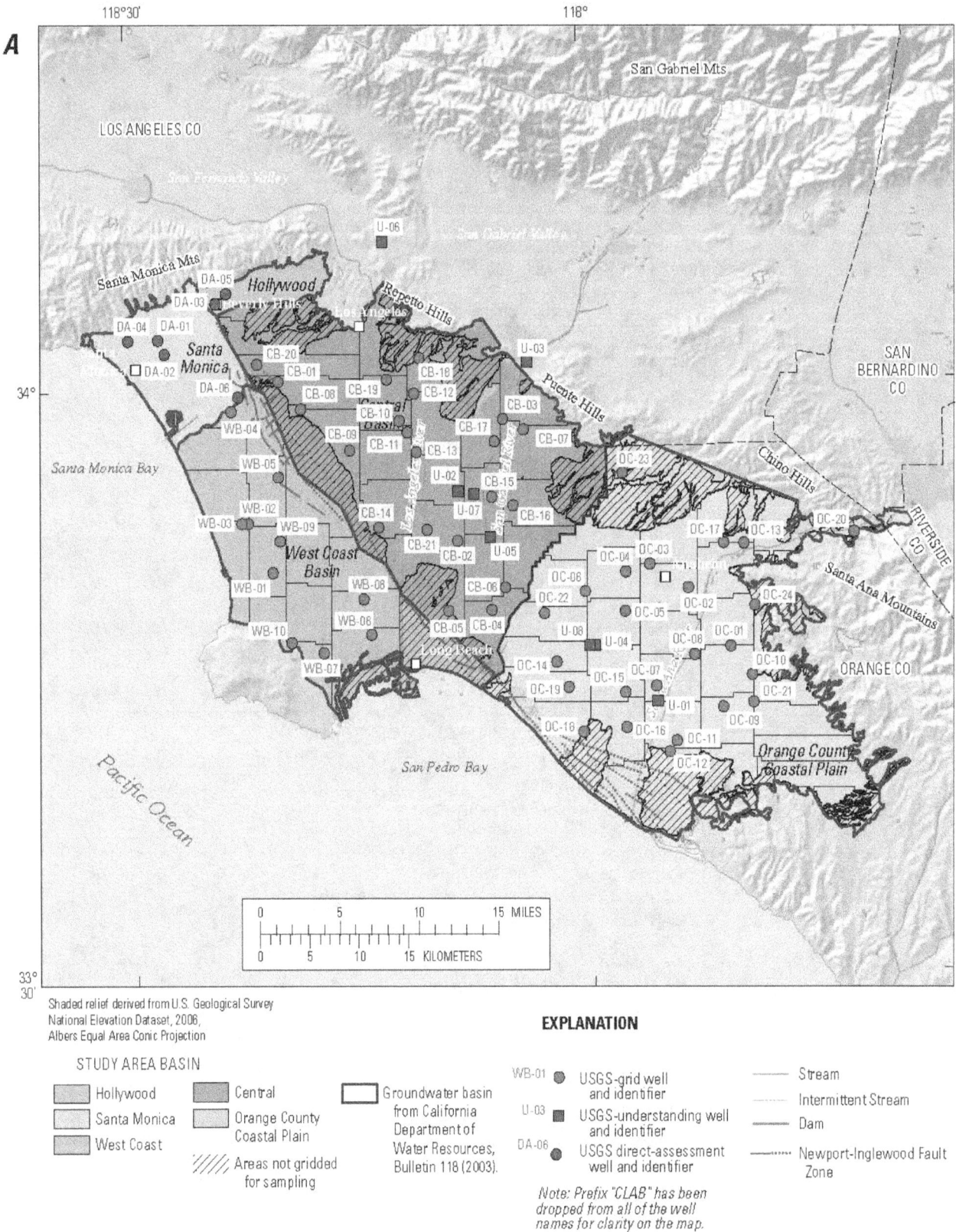

Figure A1. Identifiers and locations of (A) grid, understanding, and direct-assessment wells sampled by the USGS during June to November 2006, (B) grid wells sampled by the USGS for which data for inorganic constituents from the California Department of Public Health (CDPH) database were used, and (C) grid wells not sampled by the USGS for which data for inorganic constituents from the CDPH database were used, Coastal Los Angeles Basin study unit, 2006, California GAMA Priority Basin Project.

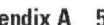

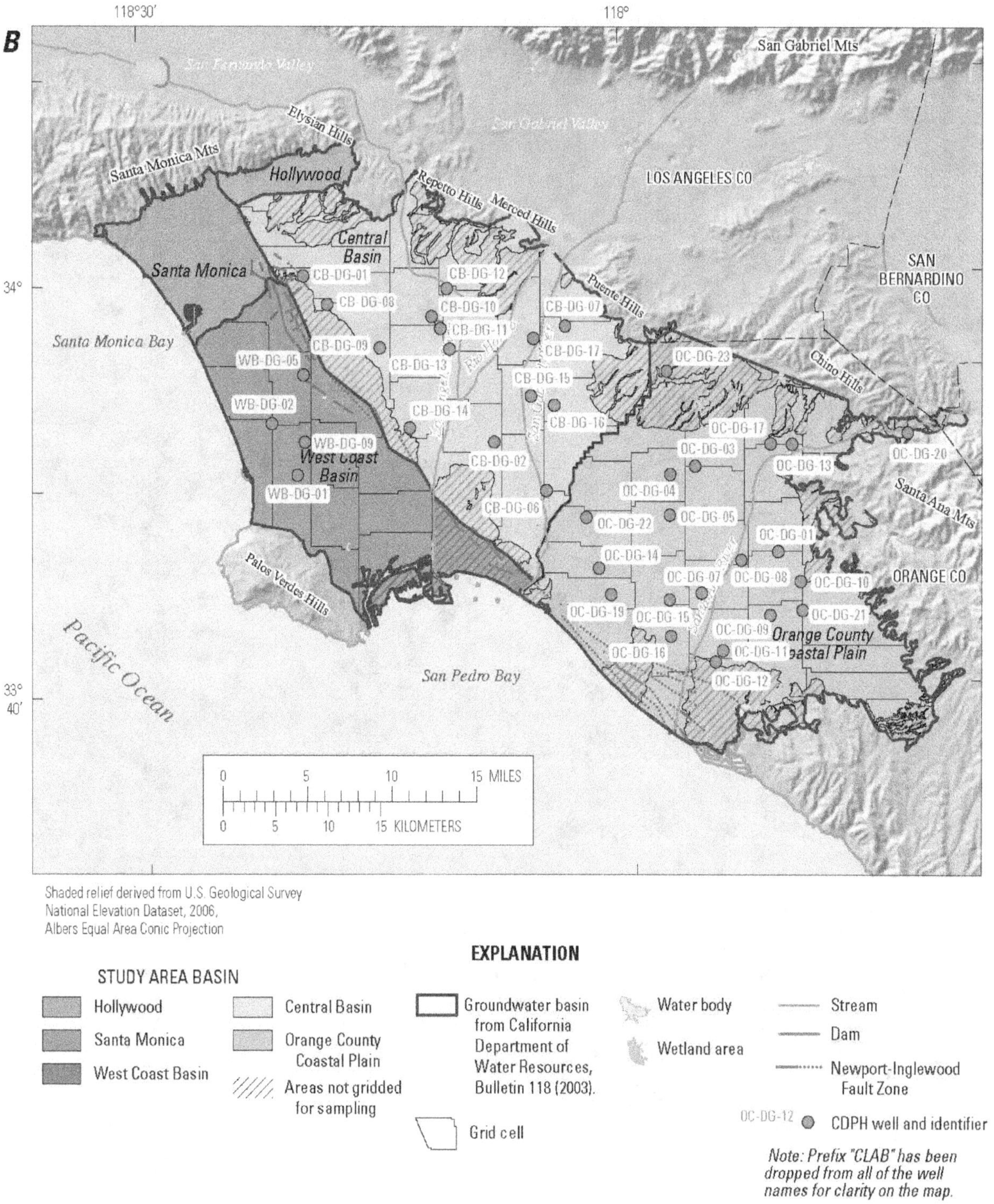
B
118°30′
118°
34°
33° 40′
San Fernando Valley
San Gabriel Mts
San Gabriel Valley
Elysian Hills
Santa Monica Mts
Hollywood
Repetto Hills
Merced Hills
Puente Hills
Chino Hills
Santa Ana Mts
Palos Verdes Hills
LOS ANGELES CO
SAN BERNARDINO CO
ORANGE CO
Central Basin
Santa Monica
West Coast Basin
Orange County Coastal Plain
Santa Monica Bay
San Pedro Bay
Pacific Ocean
CB-DG-01
CB-DG-02
CB-DG-06
CB-DG-07
CB-DG-08
CB-DG-09
CB-DG-10
CB-DG-11
CB-DG-12
CB-DG-13
CB-DG-14
CB-DG-15
CB-DG-16
CB-DG-17
WB-DG-01
WB-DG-02
WB-DG-05
WB-DG-09
OC-DG-01
OC-DG-03
OC-DG-04
OC-DG-05
OC-DG-07
OC-DG-08
OC-DG-09
OC-DG-10
OC-DG-11
OC-DG-12
OC-DG-13
OC-DG-14
OC-DG-15
OC-DG-16
OC-DG-17
OC-DG-19
OC-DG-20
OC-DG-21
OC-DG-22
OC-DG-23
0 5 10 15 MILES
0 5 10 15 KILOMETERS
Shaded relief derived from U.S. Geological Survey
National Elevation Dataset, 2006,
Albers Equal Area Conic Projection
EXPLANATION
STUDY AREA BASIN
Hollywood
Santa Monica
West Coast Basin
Central Basin
Orange County Coastal Plain
Areas not gridded for sampling
Groundwater basin from California Department of Water Resources, Bulletin 118 (2003).
Grid cell
Water body
Wetland area
Stream
Dam
Newport-Inglewood Fault Zone
OC-DG-12 CDPH well and identifier
Note: Prefix "CLAB" has been dropped from all of the well names for clarity on the map.

Figure A1.—Continued

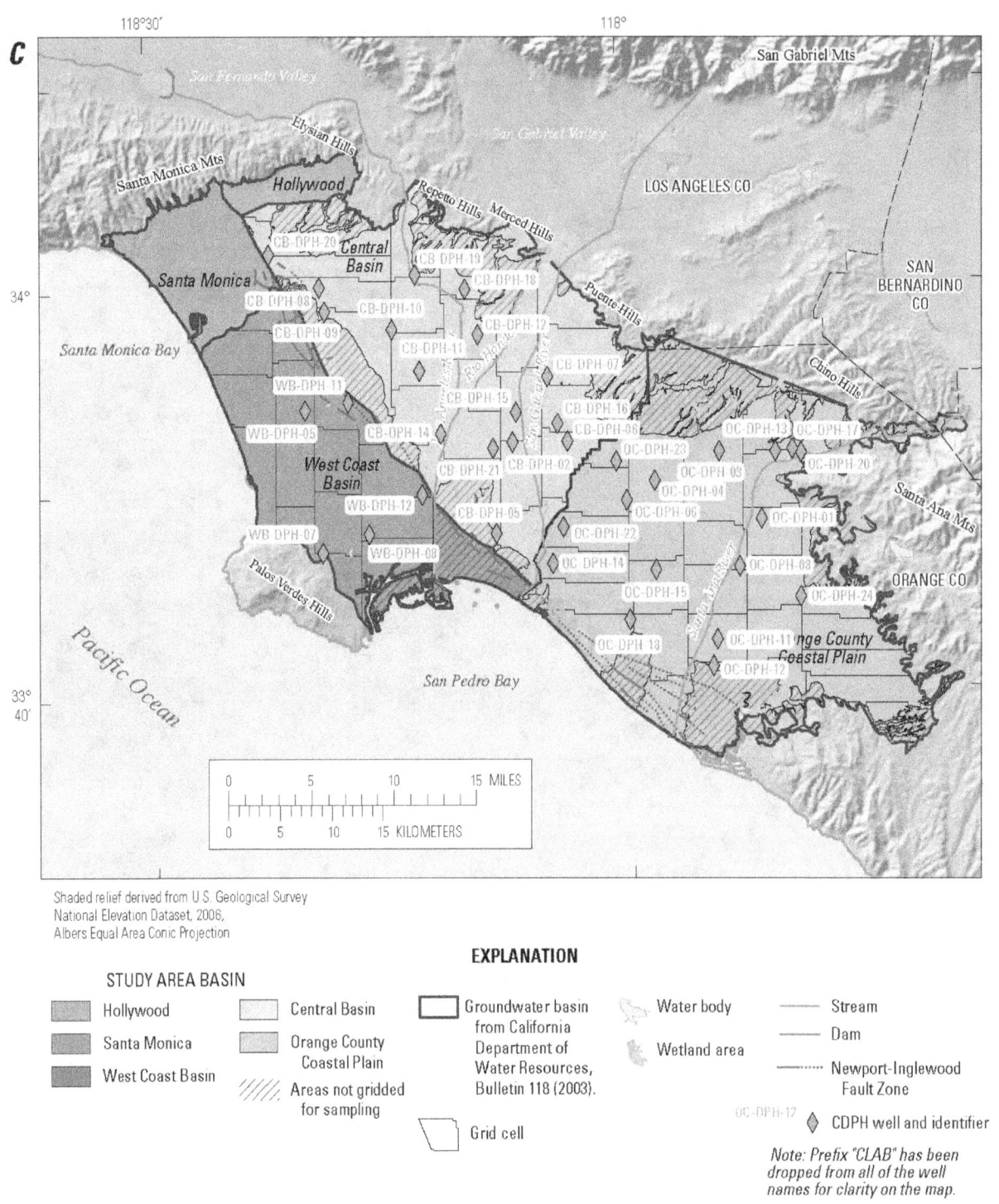

Shaded relief derived from U.S. Geological Survey
National Elevation Dataset, 2006,
Albers Equal Area Conic Projection

EXPLANATION

STUDY AREA BASIN

Hollywood

Santa Monica

West Coast Basin

Central Basin

Orange County
Coastal Plain

Areas not gridded
for sampling

Groundwater basin
from California
Department of
Water Resources,
Bulletin 118 (2003).

Grid cell

Water body

Wetland area

Stream

Dam

Newport-Inglewood
Fault Zone

OC-DPH-12   CDPH well and identifier

*Note: Prefix "CLAB" has been
dropped from all of the well
names for clarity on the map.*

Figure A1.—Continued

Appendix B. Estimation of Aquifer-Scale Proportions

The status assessment is intended to characterize the quality of groundwater resources in the primary aquifer system of the CLAB study unit. The primary aquifer system is defined by the screened or perforated depth intervals of the wells listed in the CDPH database. The use of the term "primary aquifer system" does not imply that there exists a discrete aquifer unit. In most groundwater basins, municipal and community supply wells generally are screened or perforated at greater depths than domestic wells. Thus, because domestic wells are not listed in the CDPH database, the primary aquifer system generally corresponds to the portion of the aquifer system tapped by municipal and community supply wells. Nearly all of the wells used in the status assessment are listed in the CDPH database and are therefore classified as municipal or community supply wells. However, to the extent that domestic wells are screened or perforated over the same depth intervals as the CDPH wells, the assessments presented in this report also may be applicable to the portions of the aquifer system used for domestic drinking-water supplies.

Two statistical approaches, grid-based and spatially weighted, were selected to evaluate the aquifer-scale proportions of the primary aquifer system in the CLAB study unit with high, moderate, or low relative-concentrations (concentration relative to its water-quality benchmark) of constituents (Belitz and others, 2010). The grid-based and spatially weighted estimations of aquifer-scale proportions, based on a spatially distributed grid cell network across the Central Basin, Orange County Coastal Plain, and West Basin study areas, are intended to characterize the water quality of the primary aquifer system, or at depths from which drinking water is usually drawn. These approaches assign weights to wells based on a single value per cell (grid-based) or the number of values per cells (spatially weighted). Raw detection frequencies, derived from the percentage of the total number of wells with high or moderate relative-concentrations, also were calculated for individual constituents, but were not used for estimating aquifer-scale proportion because this method creates spatial bias towards regions with large numbers of wells.

- Grid-based: One value in each grid cell, a "grid well," was randomly selected to represent the primary aquifer system (Belitz and others, 2010). For organic constituents, the one value in each grid cell was obtained from the samples analyzed by USGS-GAMA from 55 grid wells. For inorganic constituents, the one value in each grid cell was obtained from the samples analyzed by USGS-GAMA from 8 of the grid wells and from data in the CDPH database for 38 of the grid wells sampled by USGS-GAMA and 39 other wells. The relative-concentration for each constituent (concentration relative to its benchmark)

was then evaluated for each grid cell. The proportion of the primary aquifer system with high relative-concentrations was calculated by dividing the number of cells with concentrations greater than the benchmark (relative-concentration greater than 1) by the total number of grid cells with data for that constituent. Proportions containing moderate and low relative-concentrations were calculated similarly. Confidence intervals for grid-based aquifer proportions were computed using the Jeffreys interval for the binomial distribution (Brown and others, 2001). The grid-based estimate is spatially unbiased; however, the grid-based approach may not identify constituents that exist at high concentrations in small proportions of the primary aquifer system.

- Spatially weighted: The spatially weighted approach used data from the 55 grid wells and 7 of the understanding wells sampled and analyzed by USGS-GAMA in 2006, and data compiled in the CDPH database for samples collected between June 4, 2003, and June 4, 2006 (the most recent analysis was used for each constituent at each well). The spatially weighted approach uses all of the wells in each cell to calculate the high, moderate, and low relative-concentrations for the cell. The high, moderate, and low aquifer-scale proportions are calculated for each cell, and then the proportions are averaged for all of the cells with data for the constituent (Isaaks and Srivastava, 1989). The resulting proportions are spatially unbiased. Confidence intervals for spatially weighted estimates of aquifer-scale proportion are not described in this report.

The raw detection frequency approach merely is the percentage (frequency) of wells within the study unit with relative-concentrations. It was calculated by considering all of the available data in the period from June 4, 2003, to June 4, 2006, for the CDPH well data (the most recent analysis per well for all wells), the USGS-grid well data, and USGS-understanding wells. However, this approach is not spatially unbiased because the CDPH and USGS-understanding wells are not uniformly distributed. Consequently, high (or low values) for wells clustered in a particular area represent a small part of the primary aquifer system and could be given a disproportionately high (or low) weight compared to that given by spatially unbiased approaches. Raw detection frequencies of high relative-concentrations are provided to identify constituents for discussion in this report, but were not used to assess aquifer-scale proportions.

Appendix C. Ancillary Datasets

Land-Use Classification

Land use was classified using an "enhanced" version of the satellite-derived [98-ft (30-m) pixel resolution] USGS National Land Cover Dataset (Nakagaki and others, 2007). This dataset has been used in previous national and regional studies relating land use to water quality (Gilliom and others, 2006; Zogorski and others, 2006). The dataset characterizes land cover during the early 1990s. The imagery was classified into 25 land-cover classifications (Nakagaki and Wolock, 2005). These 25 land-cover classifications were aggregated into 3 principal land-use classes for the purpose of general categorization of land use: urban, agricultural, and natural. Average land use (proportions of urban, agricultural, and natural) for the study unit, for each study area, and for areas within a radius of 1,640 ft (500 m) surrounding each well (table C1) were calculated using ArcGIS (version 9.2) (Johnson and Belitz, 2009).

Well Construction Information

Well construction data primarily were obtained from drillers' logs filed with CDWR. In some cases, well construction data were obtained from ancillary records of well owners or the USGS National Water Information database. Well depths and depths to the tops and bottoms of the screened or perforated intervals for wells sampled by USGS-GAMA are listed in table C1. Well identification verification procedures are described by Mathany and others (2008). Well construction data were not available for wells in the CDPH database.

Classification of Geochemical Condition

Geochemical conditions were described by oxidation-reduction (redox) characteristics and pH. Redox conditions influence the mobility of many organic and inorganic constituents (McMahon and Chapelle, 2008). Along groundwater flow paths, redox conditions commonly proceed along a well-documented sequence of terminal electron acceptor processes (TEAPs); one TEAP typically dominates at a particular time and aquifer location (Chapelle and others, 1995; Chapelle, 2001). The predominant TEAPs are oxygen-reduction, nitrate-reduction, manganese-reduction, iron-reduction, sulfate-reduction, and methanogenesis. Redox conditions were assigned to each sample collected by USGS-GAMA using a modified version of the classification scheme of McMahon and Chapelle (2008) and (Jurgens and others, 2009). Dissolved oxygen (DO) data were collected by USGS-GAMA at all sites. Samples with DO > 0.5 mg/L were classified as oxic, and samples with DO ≤ 0.5 mg/L were classified as anoxic (table C2). The anoxic samples were further classified according to the TEAPs inferred from data for nitrate, manganese, and iron. Data for these constituents were obtained from USGS-GAMA where available and from the CDPH database ("DG" CDPH-grid wells). Inorganic constituent data were not available for all anoxic samples.

Measurements of pH were made by USGS-GAMA at 39 of the 69 wells sampled for this study (table C2). Values of pH were obtained from the CDPH database for 26 of the other wells.

Groundwater Age Classification

Groundwater age data and classifications are listed in table C3. Groundwater dating techniques indicate the time since the groundwater was last in contact with the atmosphere. The techniques used in this report to estimate groundwater residence times or 'age' were those based on tritium (for example: Tolstikhin and Kamensky, 1969; Torgersen and others, 1979) and carbon-14 activities (for example: Vogel and Ehhalt, 1963; Plummer and others, 1993).

Tritium is a short-lived radioactive isotope of hydrogen with a half-life of 12.32 years (Lucas and Unterweger, 2000). Tritium is produced naturally in the atmosphere from the interaction of cosmogenic radiation with nitrogen (Craig and Lal, 1961), by above-ground nuclear explosions, and by the operation of nuclear reactors. Tritium enters the hydrological cycle following oxidation to tritiated water. Tritium values in precipitation under natural conditions would be about 3 to 15 TU (Craig and Lal, 1961; Clark and Fritz, 1997). Above-ground nuclear explosions resulted in a large increase in tritium values in precipitation, beginning in about 1952 and peaking in 1963 at values over 1,000 TU in the northern hemisphere (Michel, 1989). Radioactive decay over a period of 50 years would decrease tritium values of 10 TU to 0.6 TU.

Previous investigations have used a range of tritium values from 0.3 to 1.0 TU as thresholds for indicating presence of water that has exchanged with the atmosphere since 1952 (Michel, 1989; Plummer and others, 1993; Michel and Schroeder, 1994; Clark and Fritz, 1997; Manning and others, 2005). For samples collected for the CLAB study unit in 2006, tritium values greater than a threshold of 1.0 TU were defined as indicating presence of groundwater recharged since 1952. By using a tritium value of 1.0 TU for the threshold in this study, the age classification scheme allows a slightly larger fraction of modern groundwater to be classified as pre-modern than if a lower threshold were used. A lower threshold for tritium would result in fewer samples classified as pre-modern than mixed, when carbon-14 would suggest that they were primarily pre-modern. This higher threshold was considered more appropriate for this study because many of the wells were production wells with long screens and mixing of waters of different ages is likely to occur.

Carbon-14 (^{14}C) is a widely used chronometer based on the radiocarbon content of organic and inorganic carbon. Dissolved inorganic carbon species, carbonic acid, bicarbonate, and carbonate typically are used for ^{14}C dating of groundwater. ^{14}C is formed in the atmosphere by the interaction of cosmic-ray neutrons with nitrogen and, to a lesser degree, with oxygen and carbon. ^{14}C is incorporated into carbon dioxide and mixed throughout the atmosphere. The carbon dioxide enters the hydrologic cycle because it dissolves in precipitation and surface water in contact with the atmosphere. ^{14}C activity in groundwater, expressed as percent modern carbon, reflects the time since groundwater was last exposed to the atmospheric ^{14}C source. ^{14}C has a half-life of 5,730 years and can be used to estimate groundwater ages ranging from 1,000 to approximately 30,000 years before present.

The ^{14}C age (residence time, presented in years) is calculated on the basis of the decrease in ^{14}C activity as a result of radioactive decay since groundwater recharge, relative to an assumed initial ^{14}C concentration (Clark and Fritz, 1997). An average initial ^{14}C activity of 100 percent modern carbon (pmc) is assumed for this study, with estimated errors on calculated groundwater ages up to ± 20%. Calculated ^{14}C ages in this study are referred to as "uncorrected" because they have not been adjusted to consider exchanges with sedimentary sources of carbon (Fontes and Garnier, 1979). Groundwater with a ^{14}C activity of >88 pmc is reported as having an age of <1,000 years; no attempt is made to refine ^{14}C ages <1,000 years. Measured values of percent modern carbon can be >100 pmc because the definition of the ^{14}C activity in "modern" carbon does not include the excess ^{14}C produced in the atmosphere by above-ground nuclear weapons testing. For the CLAB study unit, ^{14}C activity <90 pmc was defined as indicative of presence of groundwater recharged before the modern era. The threshold value of 90 pmc was selected because all groundwater samples with tritium < 1.0 TU also had ^{14}C < 90 pmc.

In this study, the age distributions of samples are classified as pre-modern, modern, and mixed. Groundwater with tritium activity less than 1 TU, and ^{14}C less than 90 pmc is designated as pre-modern, defined as having been recharged before 1952. Groundwater with tritium activities greater than 1 TU and ^{14}C greater than 90 pmc is designated as modern, defined as having been recharged after 1952. Samples with pre-modern and modern components are designated as mixed groundwater, which includes substantial fractions of old and young waters. In reality, pre-modern groundwater could contain very small fractions of modern groundwater, and modern groundwater could contain small fractions of pre-modern groundwater. Tritium concentrations, uncorrected ^{14}C age, and sample age classifications are reported in table C3. Although more sophisticated lumped parameter models used for analyzing age distributions that incorporate mixing are available (for example, Cook and Böhlke, 2000), use of these alternative models to characterize age mixtures was beyond the scope of this report. Rather, classification into modern (recharged after 1952), mixed, and pre-modern (recharged before 1952) categories was sufficient to provide an appropriate and useful characterization for the purposes of examining groundwater quality.

Table C1.   Well construction information, land-use data, and other ancillary data for wells, Coastal Los Angeles Basin study unit, 2006, California GAMA Priority Basin Project.

[Abbreviations: ft, feet; m, meter; tanks/km^2, tanks per square kilometer; LSD, land surface datum; LUFT, leaking (or formerly leaking) underground fuel tank; na, not available]

USGS-GAMA well identification number	Well construction information			Land use within 500 m of well[1]			LUFT density[2] (tanks/km^2)	Septic tank density[3] (tanks/km^2)	Aridity index[4] (dimensionless)
	Well depth (ft below LSD)	Top of openings (ft below LSD)	Bottom of openings (ft below LSD)	Agricultural (percent)	Natural (percent)	Urban (percent)			
Central Basin study area wells									
CLABCB-01	456	200	456	0.1	2.7	97	0.65	0.12	0.27
CLABCB-02	674	602	644	0	2.5	97	1.06	12.2	0.24
CLABCB-03	520	242	446	0.9	3.0	96	1.52	0.01	0.26
CLABCB-04	838	507	838	0	0.5	100	2.73	0	0.22
CLABCB-05	910	300	898	0.7	27	73	1.34	0	0.23
CLABCB-06	1,680	605	1,640	0	40	60	0.53	0	0.23
CLABCB-07	900	200	900	0.1	11	89	5.91	0	0.26
CLABCB-08	1,096	420	1,076	0	1.7	98	1.01	0	0.26
CLABCB-09	1,428	408	1,400	0	3.2	97	0.68	5.28	0.26
CLABCB-10	1,500	500	1,500	0	2.2	98	3.03	28.2	0.26
CLABCB-11	1,504	500	1,504	0.1	1.6	98	5.01	4.81	0.26
CLABCB-12	1,182	475	1,094	0	9.6	90	3.23	62.4	0.26
CLABCB-13	746	610	746	0	17	83	4.85	18.3	0.27
CLABCB-14	501	451	501	0	4.7	95	1.48	10.4	0.25
CLABCB-15	688	626	688	0	3.7	96	0.53	0.61	0.26
CLABCB-16	736	684	718	0	0.1	100	2.45	0	0.25
CLABCB-17	627	277	584	0	0.3	100	0.96	0.50	0.26
CLABCB-18	400	na	na	0.1	4.7	95	6.00	8.38	0.27
CLABCB-19	1,010	617	973	0.1	6.2	94	1.34	0	0.26
CLABCB-20	502	331	338	0	3.0	97	3.95	6.67	0.27
CLABCB-21	660	640	660	0	8.2	92	2.37	21.7	0.25
Orange County Coastal Plain study area wells									
CLABOC-01	1,300	399	1,270	3.3	19	78	0.69	12.7	0.25
CLABOC-02	1,550	599	1,530	0	3.9	96	1.22	0	0.24
CLABOC-03	450	309	425	0	1.8	98	1.69	0	0.24
CLABOC-04	1,420	482	1,375	0	12	88	3.01	1.45	0.24
CLABOC-05	1,240	505	1,220	0	0	100	1.49	0	0.23
CLABOC-06	1,300	540	1,280	1.6	8.8	90	1.28	9.80	0.23
CLABOC-07	972	390	940	0	1.0	99	3.00	28.8	0.24
CLABOC-08	1,310	570	1,290	0.3	1.5	98	1.38	0	0.25
CLABOC-09	1,152	330	1,140	0	4.2	96	6.17	0	0.24
CLABOC-10	1,180	560	1,160	0.1	3.8	96	1.48	19.0	0.24
CLABOC-11	600	305	580	0	25	75	3.27	0	0.23
CLABOC-12	1,060	310	1,025	0.1	38	62	1.85	0	0.23
CLABOC-13	420	90	406	0	13	87	4.51	2.68	0.25
CLABOC-14	880	374	860	0	2.4	98	1.47	0	0.22
CLABOC-15	1,135	345	1,125	0.1	3.4	96	1.56	0	0.23
CLABOC-16	366	201	356	0.2	7.6	92	3.87	0	0.23
CLABOC-17	1,230	530	1,210	0	51	49	1.47	0	0.24
CLABOC-18	486	342	486	0.1	26	74	0.97	1.80	0.22
CLABOC-19	306	265	291	3.8	10	86	2.02	0	0.22
CLABOC-20	98	60	84	3.6	61	36	0.08	0.20	0.29
CLABOC-21	998	397	995	0	3.8	96	1.28	0	0.23
CLABOC-22	260	na	na	0	36	64	3.89	0.27	0.22
CLABOC-23	970	460	950	0	7.6	92	2.85	0.65	0.26
CLABOC-24	604	256	584	0	2.7	97	0.81	6.44	0.26

Table C1. Well construction information, land-use data, and other ancillary data for wells, Coastal Los Angeles Basin study unit, 2006, California GAMA Priority Basin Project.—Continued

[Abbreviations: ft, feet; m, meter; tanks/km², tanks per square kilometer; LSD, land surface datum; LUFT, leaking (or formerly leaking) underground fuel tank; na, not available]

USGS-GAMA well identification number	Well construction information			Land use within 500 m of well[1]			LUFT density[2] (tanks/km²)	Septic tank density[3] (tanks/km²)	Aridity index[4] (dimensionless)
	Well depth (ft below LSD)	Top of openings (ft below LSD)	Bottom of openings (ft below LSD)	Agricultural (percent)	Natural (percent)	Urban (percent)			
West Coast Basin study area wells									
CLABWB-01	445	310	425	0.2	26	74	2.79	2.56	0.25
CLABWB-02	490	210	420	0	1.7	98	1.61	0	0.25
CLABWB-03	445	215	425	0.1	1.9	98	8.15	0	0.25
CLABWB-04	620	200	600	0	12	88	1.01	0	0.26
CLABWB-05	800	340	730	0	3.7	96	1.69	3.16	0.25
CLABWB-06	930	480	910	0.2	6.5	93	0.65	1.73	0.23
CLABWB-07	822	630	800	0	26	74	0.70	0	0.25
CLABWB-08	780	450	750	0	20	80	2.30	0.71	0.24
CLABWB-09	810	200	786	0	4.9	95	1.77	0	0.25
CLABWB-10	600	300	600	0	13	87	0.67	21.4	0.29
Understanding wells									
CLABU-01	844	312	844	0	0	100	2.97	4.08	0.24
CLABU-02	572	546	572	0	0.9	99	2.08	11.9	0.27
CLABU-03	414	183	386	11	33	56	1.02	0.20	0.27
CLABU-04	540	497	540	0.3	4.2	95	5.75	0	0.22
CLABU-05	222	212	220	0.1	16	84	1.46	6.11	0.24
CLABU-06	300	62	300	0	1.7	98	1.70	19.6	0.33
CLABU-07	na	na	na	0.6	2.5	97	0.97	0	0.27
CLABU-08	120	na	na	0	15	85	3.36	0	0.22
Direct assessment wells									
CLABDA-01	250	120	220	0	1.5	99	1.92	4.94	0.29
CLABDA-02	550	210	530	0	3.9	96	2.28	5.68	0.28
CLABDA-03	665	360	630	0	1.9	98	5.93	0.11	0.29
CLABDA-04	282	151	250	0	18	82	0.18	1.05	0.30
CLABDA-05	740	398	730	0	0.2	100	1.61	1.25	0.30
CLABDA-06	440	260	430	0.7	39	61	0.39	0	0.26

[1]Land-use percentages within 500-m radius of well site (Nakagaki and others, 2007; Johnson and Belitz, 2009).

[2]Leaking (or formerly leaking) underground fuel tank density within 500-m radius of well site (California State Water Resources Control Board, 2001).

[3]Septic tank density within 500-m radius of well site (U.S. Census Bureau, 1990).

[4]Aridity index is average annual precipitation divided by average annual potential evapotranspiration (UNESCO, 1979).

Table C2.   Oxidation-reduction classification, dissolved oxygen concentration, pH, and ratio of oxidized and reduced iron, Coastal Los Angeles Basin study unit, 2006, California GAMA Priority Basin Project.

[Anoxic reduction processes: none, groundwater is oxic and Mn and Fe concentrations are lower than the threshold for identifying Mn-red or Fe-red; unknown, groundwater is oxic or anoxic and there are no Mn and Fe concentration data; NO3-red, nitrate-reduction; Mn-red, manganese-reduction; Fe-red, iron-reduction. Ratio of oxidized to reduced species of iron: Fe^{+3}/Fe^{+2}, ratio of the amount of iron in the +3 oxidation state (ferric) to the amount of iron in the +2 oxidation state (ferrous). Other abbreviations: CDPH, California Department of Public Health; USGS, U.S. Geological Survey; mg/L, milligrams per liter; <, less than; >, greater than; –, concentration too low to measure ratio; nc, not collected]

Well identification number	Source of pH data	pH (standard units)	Dissolved oxygen (mg/L)	Fe^{+3}/Fe^{+2}	Source of NO$_3$, Mn, and Fe concentration data	Oxidation-reduction classification	Anoxic reduction process
Central Basin study area wells							
CLABCB-01	USGS	7.1	0.6	–	CDPH	Oxic	none
CLABCB-02	USGS	7.3	0.2	–	USGS	Anoxic	NO$_3$-red
CLABCB-03	USGS	7.1	0.8	–	USGS	Oxic	none
CLABCB-04	USGS	8.7	0.2	–	USGS	Anoxic	Suboxic
CLABCB-05	nc	nc	0.2	–	nc	Anoxic	unknown
CLABCB-06	CDPH	8.2	1.2	0.2	CDPH	Oxic	none
CLABCB-07	CDPH	7.6	1.7	–	CDPH	Oxic	none
CLABCB-08	CDPH	7.8	1.0	–	nc	Oxic	unknown
CLABCB-09	CDPH	7.9	0.2	0.6	nc	Anoxic	unknown
CLABCB-10	CDPH	7.9	0.2	1	CDPH	Anoxic	Suboxic
CLABCB-11	CDPH	7.9	0.4	–	CDPH	Anoxic	NO$_3$-red
CLABCB-12	CDPH	7.9	0.2	7	CDPH	Anoxic	Mn-red
CLABCB-13	CDPH	7.5	1.2	–	nc	Oxic	unknown
CLABCB-14	CDPH	8.2	0.2	0.1	CDPH	Anoxic	Suboxic
CLABCB-15	CDPH	7.9	1.1	–	CDPH	Oxic	none
CLABCB-16	nc	nc	3.8	2	CDPH	Oxic	Fe-red[1]
CLABCB-17	USGS	7.3	3.0	–	USGS	Oxic	none
CLABCB-18	USGS	7.7	0.2	< 0.1	nc	Anoxic	unknown
CLABCB-19	USGS	7.2	3.6	–	nc	Oxic	unknown
CLABCB-20	USGS	7.5	0.2	< 0.1	nc	Anoxic	unknown
CLABCB-21	USGS	7.8	0.2	0.4	nc	Anoxic	unknown
Orange County Coastal Plain study area wells							
CLABOC-01	USGS	7.4	1.5	nc	USGS	Oxic	none
CLABOC-02	USGS	7.5	1.1	–	USGS	Oxic	none
CLABOC-03	CDPH	8.0	5.4	–	CDPH	Oxic	none
CLABOC-04	CDPH	8.1	3.2	–	CDPH	Oxic	none
CLABOC-05	CDPH	8.2	3.5	–	CDPH	Oxic	none
CLABOC-06	nc	nc	4.0	–	nc	Oxic	unknown
CLABOC-07	CDPH	7.2	4.5	–	CDPH	Oxic	none
CLABOC-08	CDPH	7.5	4.6	–	CDPH	Oxic	none
CLABOC-09	CDPH	8.2	2.9	–	CDPH	Oxic	none
CLABOC-10	CDPH	8.1	2.6	–	CDPH	Oxic	none
CLABOC-11	CDPH	8.1	0.5	–	CDPH	Anoxic	Suboxic
CLABOC-12	CDPH	8.7	0.5	–	nc	Anoxic	unknown
CLABOC-13	CDPH	8.0	0.3	–	CDPH	Anoxic	NO$_3$-red
CLABOC-14	CDPH	8.3	0.3	0.4	CDPH	Anoxic	Suboxic
CLABOC-15	CDPH	8.1	3.9	–	CDPH	Oxic	none
CLABOC-16	CDPH	8.1	0.6	–	CDPH	Oxic	none
CLABOC-17	CDPH	8.0	2.5	–	CDPH	Oxic	none
CLABOC-18	USGS	8.8	0.2	–	nc	Anoxic	unknown
CLABOC-19	USGS	7.5	0.2	–	CDPH	Anoxic	Suboxic
CLABOC-20	USGS	7.2	0.2	–	USGS	Anoxic	NO$_3$-red, Mn-red
CLABOC-21	USGS	7.4	3.4	5	CDPH	Oxic	none
CLABOC-22	USGS	7.9	7.0	0.4	CDPH	Oxic	Fe-red[1]
CLABOC-23	USGS	7.7	0.3	< 0.1	CDPH	Anoxic	Fe-red
CLABOC-24	USGS	7.2	2.5	> 10	nc	Oxic	unknown

Table C2. Oxidation-reduction classification, dissolved oxygen concentration, pH, and ratio of oxidized and reduced iron, Coastal Los Angeles Basin study unit, 2006, California GAMA Priority Basin Project.—Continued

[**Anoxic reduction processes**: none, groundwater is oxic, and Mn and Fe concentrations are lower than the threshold for identifying Mn-red or Fe-red; unknown, groundwater is oxic or anoxic, and there are no Mn and Fe concentration data; NO_3-red, nitrate-reduction; Mn-red, manganese-reduction; Fe-red, iron-reduction. **Ratio of oxidized to reduced species of iron**: Fe^{+3}/Fe^{+2}, ratio of the amount of iron in the +3 oxidation state (ferric) to the amount in the +2 oxidation state (ferrous). Other abbreviations: CDPH, California Department of Public Health; USGS, U.S. Geological Survey; mg/L, milligrams per liter; <, less than; >, greater than; –, concentration too low to measure ratio; nc, not collected]

Well identification number	Source of pH data	pH (standard units)	Dissolved oxygen (mg/L)	Fe^{+3}/Fe^{+2}	Source of NO_3, Mn, and Fe concentration data	Oxidation-reduction classification	Anoxic reduction process
West Coast Basin study area wells							
CLABWB-01	CDPH	7.8	0.2	0.2	CDPH	Anoxic	Mn-red
CLABWB-02	CDPH	7.6	2.1	0.3	CDPH	Oxic	none
CLABWB-03	nc	nc	0.3	<0.1	nc	Anoxic	unknown
CLABWB-04	USGS	7.6	0.2	<0.1	USGS	Anoxic	Mn-red, Fe-red
CLABWB-05	USGS	8.0	0.2	0.1	CDPH	Anoxic	Fe-red
CLABWB-06	USGS	8.2	0.2	<0.1	nc	Anoxic	unknown
CLABWB-07	USGS	8.0	0.2	<0.1	nc	Anoxic	unknown
CLABWB-08	USGS	8.2	3.6	0.2	nc	Oxic	unknown
CLABWB-09	USGS	6.5	0.2	<0.1	CDPH	Anoxic	Suboxic
CLABWB-10	USGS	7.4	0.2	<0.1	nc	Anoxic	unknown
Understanding wells							
CLABU-01	USGS	7.3	2.8	nc	USGS	Oxic	none
CLABU-02	USGS	7.2	0.2	–	USGS	Anoxic	NO_3-red
CLABU-03	USGS	7.3	0.8	–	USGS	Oxic	none
CLABU-04	USGS	7.5	3.7	–	nc	Oxic	unknown
CLABU-05	USGS	7.5	0.3	0.2	nc	Anoxic	unknown
CLABU-06	USGS	6.9	2.1	–	USGS	Oxic	none
CLABU-07	USGS	7.2	0.2	0.8	USGS	Anoxic	Mn-red
CLABU-08	USGS	7.4	1.3	6	nc	Oxic	unknown
Direct-assessment wells							
CLABDA-01	USGS	6.5	1.8	0.2	USGS	Oxic	none
CLABDA-02	USGS	6.7	2.5	1	USGS	Oxic	none
CLABDA-03	USGS	7.4	0.2	<0.1	USGS	Anoxic	Suboxic
CLABDA-04	USGS	6.5	2.9	1	USGS	Oxic	none
CLABDA-05	USGS	7.8	0.2	<0.1	USGS	Anoxic	Suboxic
CLABDA-06	USGS	7.2	0.2	<0.1	USGS	Anoxic	Mn-red, Fe-red

[1] Iron concentration from CDPH data was greater than the threshold for identifying iron-reducing conditions; however, the Fe^{+3}/Fe^{+2} ratio was higher than expected for iron-reducing conditions. The difference likely reflects that the sampling dates for USGS-GAMA and CDPH were different, and water quality may vary between sampling events. The samples were classified as oxic on the basis of the dissolved oxygen data alone.

Table C3.   Tritium and carbon-14 data and groundwater age classifications, Coastal Los Angeles Basin study unit, 2006, California GAMA Priority Basin Project.

[Groundwater age classifications were based on tritium and carbon-14 data. Groundwater with tritium < 1 TU was defined as pre-modern, recharged before 1952. Groundwater with tritium > 1 TU and percent modern carbon > 88 was defined as modern, recharged after 1952. Groundwater with tritium > 1 TU and percent modern carbon < 88 was defined as mixed, containing components recharged before and after 1952. In the absence of carbon-14 data, groundwater with tritium > 1 TU was defined as modern or mixed. Abbreviations: ^{14}C, carbon-14; TU, tritium units; nc, not collected; <, less than; >, greater than]

GAMA_ID	Tritium (TU)	Percent modern carbon	Uncorrected ^{14}C age (years)	Age classification
Central Basin study area wells				
CLABCB-01	0.11	nc	nc	Pre-modern
CLABCB-02	5.78	96	<1,000	Modern
CLABCB-03	6.25	98	<1,000	Modern
CLABCB-04	0.18	29	9,950	Pre-modern
CLABCB-05	0.72	nc	nc	Pre-modern
CLABCB-06	0.54	nc	nc	Pre-modern
CLABCB-07	4.75	nc	nc	Modern or Mixed
CLABCB-08	−0.14	nc	nc	Pre-modern
CLABCB-09	0.20	nc	nc	Pre-modern
CLABCB-10	1.21	nc	nc	Modern or Mixed
CLABCB-11	8.27	nc	nc	Modern or Mixed
CLABCB-12	7.72	nc	nc	Modern or Mixed
CLABCB-13	6.09	nc	nc	Modern or Mixed
CLABCB-14	0.10	nc	nc	Pre-modern
CLABCB-15	7.79	nc	nc	Modern or Mixed
CLABCB-16	10.83	nc	nc	Modern or Mixed
CLABCB-17	9.12	nc	nc	Modern or Mixed
CLABCB-18	0.48	nc	nc	Pre-modern
CLABCB-19	0.09	nc	nc	Pre-modern
CLABCB-20	0.88	61	3,970	Pre-modern
CLABCB-21	0.48	nc	nc	Pre-modern
Orange County Coastal Plain study area wells				
CLABOC-01	3.53	nc	nc	Modern or Mixed
CLABOC-02	4.32	108	<1,000	Modern
CLABOC-03	8.52	nc	nc	Modern or Mixed
CLABOC-04	5.90	nc	nc	Modern or Mixed
CLABOC-05	9.52	nc	nc	Modern or Mixed
CLABOC-06	7.04	nc	nc	Modern or Mixed
CLABOC-07	0.27	nc	nc	Pre-modern
CLABOC-08	6.32	nc	nc	Modern or Mixed
CLABOC-09	0.71	nc	nc	Pre-modern
CLABOC-10	1.27	nc	nc	Modern or Mixed
CLABOC-11	0.19	nc	nc	Pre-modern
CLABOC-12	0.13	nc	nc	Pre-modern
CLABOC-13	5.01	nc	nc	Modern or Mixed
CLABOC-14	0.19	nc	nc	Pre-modern
CLABOC-15	0.08	nc	nc	Pre-modern
CLABOC-16	6.99	nc	nc	Modern or Mixed
CLABOC-17	6.22	nc	nc	Modern or Mixed
CLABOC-18	0.20	nc	nc	Pre-modern
CLABOC-19	0.11	nc	nc	Pre-modern
CLABOC-20	5.27	101	<1,000	Modern
CLABOC-21	1.33	nc	nc	Modern or Mixed
CLABOC-22	0.19	nc	nc	Pre-modern
CLABOC-23	−0.31	nc	nc	Pre-modern
CLABOC-24	5.61	nc	nc	Modern or Mixed

Table C3.   Tritium and carbon-14 data and groundwater age classifications, Coastal Los Angeles Basin study unit, 2006, California GAMA Priority Basin Project.—Continued

[Groundwater age classifications were based on tritium and carbon-14 data. Groundwater with tritium < 1 TU was defined as pre-modern, recharged before 1952. Groundwater with tritium > 1 TU and percent modern carbon > 88 was defined as modern, recharged after 1952. Groundwater with tritium > 1 TU and percent modern carbon < 88 was defined as mixed, containing components recharged before and after 1952. In the absence of carbon-14 data, groundwater with tritium > 1 TU was defined as modern or mixed. Abbreviations: ^{14}C, carbon-14; TU, tritium units; nc, not collected; <, less than; >, greater than]

GAMA_ID	Tritium (TU)	Percent modern carbon	Uncorrected ^{14}C age (years)	Age classification
West Coast Basin study area wells				
CLABWB-01	45.33	nc	nc	Modern or Mixed
CLABWB-02	8.38	nc	nc	Modern or Mixed
CLABWB-03	10.46	nc	nc	Modern or Mixed
CLABWB-04	0.00	37	8,030	Pre-modern
CLABWB-05	0.13	nc	nc	Pre-modern
CLABWB-06	0.68	nc	nc	Pre-modern
CLABWB-07	0.39	nc	nc	Pre-modern
CLABWB-08	0.07	nc	nc	Pre-modern
CLABWB-09	3.81	nc	nc	Modern or Mixed
CLABWB-10	24.13	nc	nc	Modern or Mixed
Understanding wells				
CLABU-01	0.56	nc	nc	Pre-modern
CLABU-02	7.68	nc	nc	Modern or Mixed
CLABU-03	6.42	95	<1,000	Modern
CLABU-04	8.19	nc	nc	Modern or Mixed
CLABU-05	0.28	nc	nc	Pre-modern
CLABU-06	6.84	90	<1,000	Modern
CLABU-07	8.69	109	<1,000	Modern
CLABU-08	0.66	nc	nc	Pre-modern
Direct-assessment wells				
CLABDA-01	3.69	91	<1,000	Modern
CLABDA-02	3.62	nc	nc	Modern or Mixed
CLABDA-03	0.03	42	6,920	Pre-modern
CLABDA-04	4.33	96	<1,000	Modern
CLABDA-05	0.13	22	12,260	Pre-modern
CLABDA-06	5.09	nc	nc	Modern or Mixed

Appendix D. Comparison of CDPH and USGS-GAMA Data

CDPH and USGS-GAMA data were compared to assess the validity of combining data for inorganic constituents from these different sources. Concentrations of inorganic constituents (calcium, magnesium, sodium, chloride, sulfate, TDS, arsenic), which generally are prevalent at concentrations substantially above reporting levels, were compared for each well using data from both sources. Twelve wells had data for major ions or nitrate from the USGS database and the CDPH database. Wilcoxon signed-rank tests of paired analyses for these eight constituents indicated no significant differences between USGS-GAMA and CDPH data for these constituents. While differences between the paired datasets existed for some wells, most sample pairs plotted close to a 1-to-1 line (fig. D1). These plots indicated that the GAMA and CDPH inorganic data were comparable.

Major-ion data for grid wells with sufficient data (USGS and CDPH data) were plotted on a trilinear diagram (Piper, 1944) along with all CDPH major-ion data to determine whether the groundwater types in grid wells were similar to groundwater types observed in the study unit. Trilinear diagrams show the relative abundance of major cations and anions (on a charge equivalent basis) as a percentage of the total ion content of the water (fig. D2). Trilinear diagrams often are used to define groundwater type (Hem, 1985). All

major-ion data in the CDPH database with a cation/anion imbalance of less than 10 percent were retrieved and plotted on the trilinear diagrams for comparison with USGS- and CDPH-grid well data.

The ranges of water types for USGS-grid wells and other wells from the CDPH database were similar (fig. D2). In most water samples from wells, no single cation accounted for more than 60 percent of the total cations, and bicarbonate accounted for more than 60 percent of the total anions. Waters in these wells are described as *mixed cation–bicarbonate* type waters. The majority of wells contained *mixed cation–mixed anion* type waters, indicating that no single cation and no single anion accounted for more than 60 percent of the total. Three wells contain calcium/magnesium–bicarbonate waters, for which calcium plus magnesium and bicarbonate account for more than 60 percent of the cations and anions. Waters in a minority of wells are classified as sodium bicarbonate–chloride type waters, indicating that sodium and chloride accounted for more than 10 percent of the total cations and anions. The determination that the range of relative abundance of major cations and anions in grid wells is similar to the range of those in all CDPH wells indicates that the USGS-grid wells represent the types of water present in the study unit.

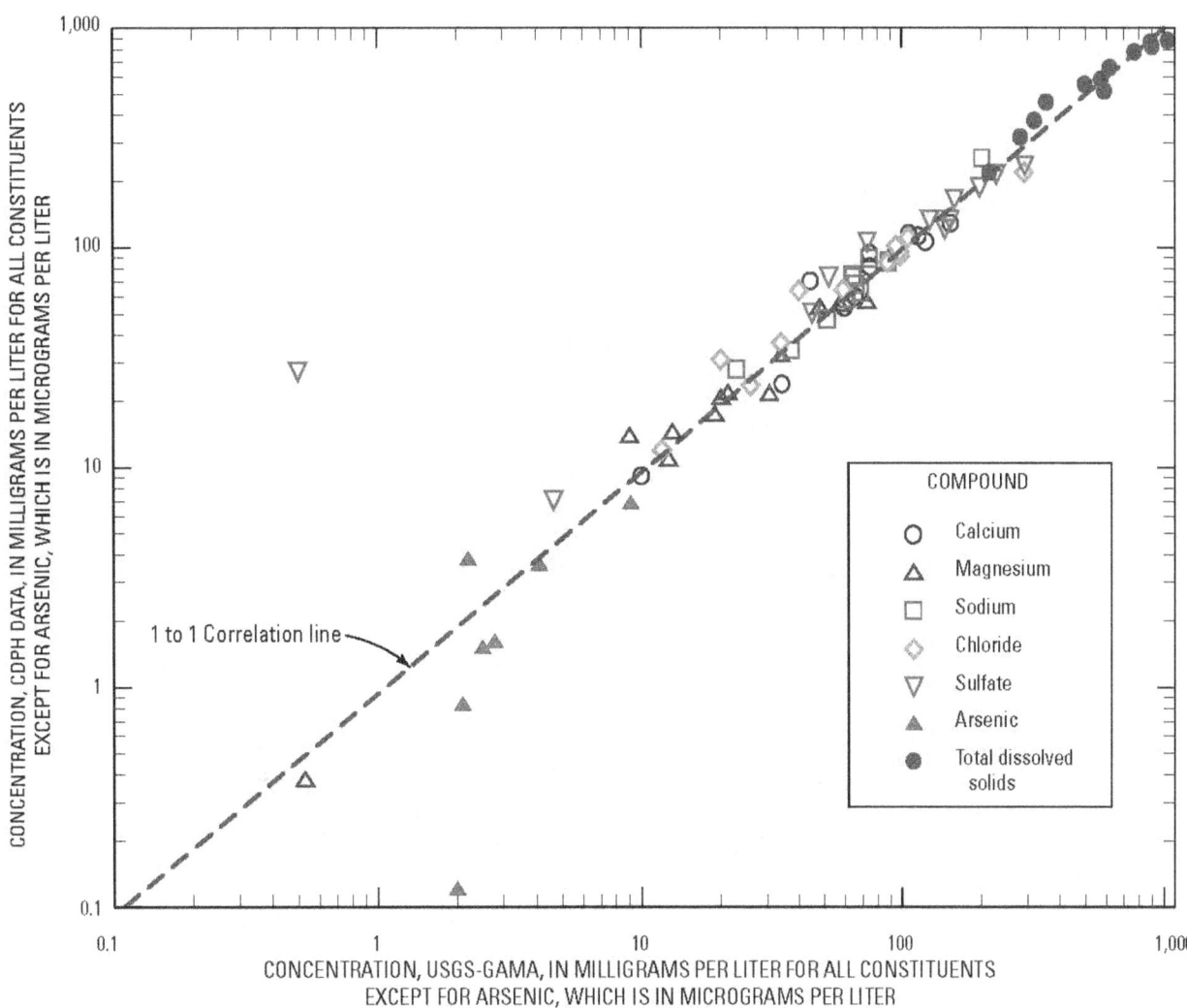

Figure D1. Paired inorganic constituent concentrations from wells sampled by the Groundwater Ambient Monitoring and Assessment (GAMA) Program June to November 2006 and prior 3-year California Department of Public Health (CDPH) data, Coastal Los Angeles Basin study unit, California GAMA Priority Basin Project.

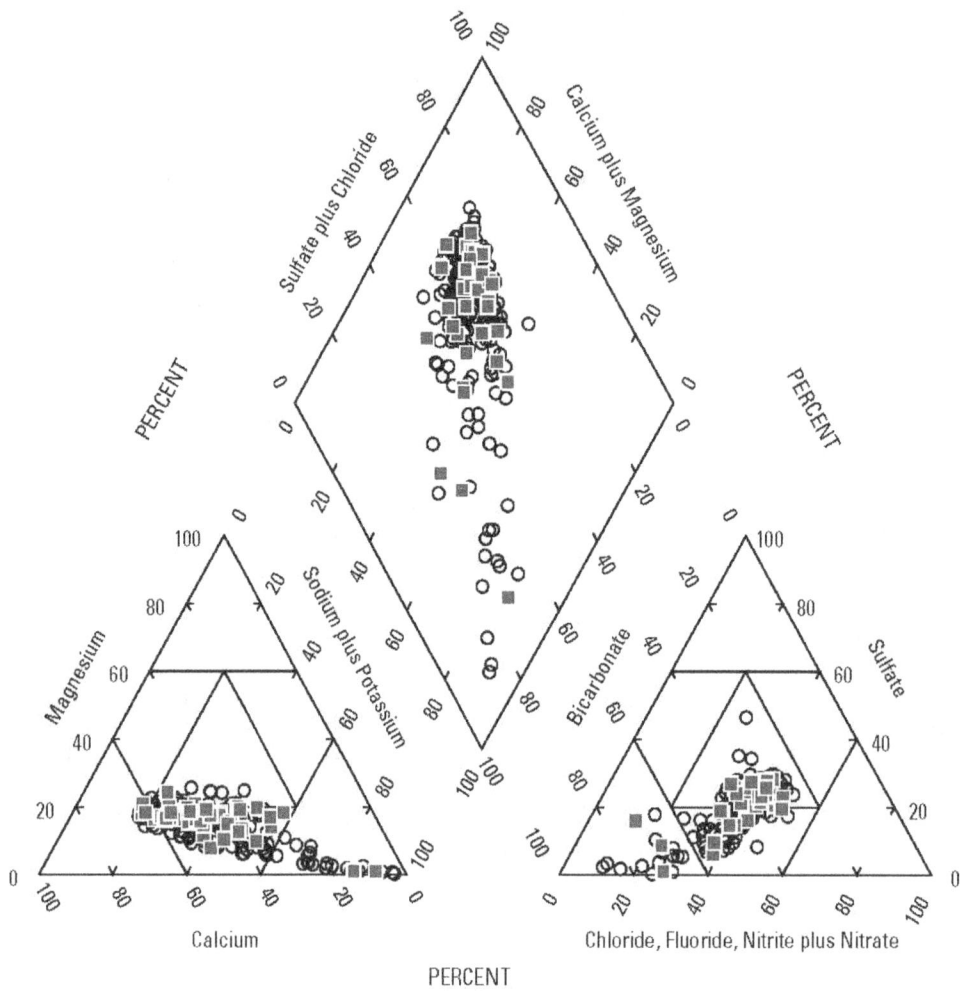

Figure D2. U.S. Geological Survey (USGS) GAMA wells and all wells in the California Department of Public Health (CDPH) database with a charge imbalance of less than 10 percent, Coastal Los Angeles Basin study unit, California GAMA Priority Basin Project.

www.ingramcontent.com/pod-product-compliance
Lightning Source LLC
Chambersburg PA
CBHW081603170526
45166CB00009B/2808